Quantum
Fluctuations

Princeton Series in Physics

edited by Arthur S. Wightman
and Philip W. Anderson

Quantum Mechanics for Hamiltonians Defined as Quadratic Forms
by Barry Simon

Lectures on Current Algebra and Its Applications
by Sam B. Treiman, Roman Jackiw, and David J. Gross

Physical Cosmology *by P.J.E. Peebles*

The Many-Worlds Interpretation of Quantum Mechanics
edited by B. S. DeWitt and N. Graham

The $P(\Phi)_2$ Euclidean (Quantum) Field Theory *by Barry Simon*

Homogeneous Relativistic Cosmologies *by Michael P. Ryan, Jr., and Lawrence C. Shepley*

Studies in Mathematical Physics: Essays in Honor of Valentine Bargmann
edited by Elliott H. Lieb, B. Simon, and A. S. Wightman

Convexity in the Theory of Lattice Gases *by Robert B. Israel*

Surprises in Theoretical Physics *by Rudolf Peierls*

The Large-Scale Structure of the Universe
by P.J.E. Peebles

Quantum Theory and Measurement *edited by John Archibald Wheeler and Wojciech Hubert Zurek*

Statistical Physics and the Atomic Theory of Matter,
from Boyle and Newton to Landau and Onsager *by Stephen G. Brush*

Quantum Fluctuations *by Edward Nelson*

Quantum Fluctuations

by

Edward Nelson

Princeton Series in Physics

Princeton University Press
Princeton, New Jersey

Published by Princeton University Press, 41 William Street,
Princeton, New Jersey 08540
IN THE UNITED KINGDOM:
Princeton University Press, Chichester, West Sussex

Library of Congress Cataloging in Publication Data will be
found on the last printed page of this book

ISBN 0-691-08378-9 (cloth) 0-691-08379-7 (paper)

Princeton University Press books are printed on acid-free paper
and meet the guidelines for permanence and durability of the
Committee on Production Guidelines for Book Longevity of the
Council on Library Resources

Printed in the United States of America

8 7 6 5 4 3 2

CONTENTS

PREFACE

These are the revised lecture notes of a course given in June, 1983 for the Troisième cycle de la physique en Suisse romande. This course was given at a time when my thinking about stochastic mechanics was in a state of flux. There are many loose ends; this is not a treatise. I publish it in the attempt to express a viewpoint about the nature of quantum fluctuations, and in the hope that others will be encouraged to work on the many mathematical problems that are left unresolved here.

I wish to express my deep thanks to P. Huguenin, W. Amrein, J.-P. Eckmann, and all of the Troisième cycle for the invitation to give this course and for the many arrangements they made, to the audience for their lively participation, to R. Chacon for an invitation to talk about stochastic mechanics in Vancouver in the summer of 1982, to W. Faris and S. Goldstein for insisting that Markovian stochastic mechanics violates locality, to A. Rechel-Cohn and L. Smolin for sharing their ideas about gravity and quantum effects, to E. Carlen for invaluable help in writing this book (finding errors, suggesting ideas, and informing me about the literature — a role reversal of the student-teacher relationship), to J. Lafferty and M. Campanino for correcting computational and conceptual mistakes, to F. Guerra for generous help and criticisms, to K. Yasue for helpful correspondence and for locating our beautiful apartment, to J.S. Bell, S. Buss, and N. Gisin for stimulating conversations, and to J.-C. Zambrini for making our stay in Geneva extremely enjoyable both scientifically and personally.

This work was partially supported by the National Science Foundation grant MCS-8101877 A02.

* * *

The thesis of this investigation is that perhaps quantum fluctuations are real, and have physical causes.

No physical system of finitely many degrees of freedom is truly isolated; it is always in interaction with a background field. Consider a system with Lagrangian

$$L = \frac{1}{2} m_{ij} p^i p^j - \phi + A_i p^i .$$

The inverse m^{ij} of the mass tensor m_{ij} is a symmetric positive definite contravariant tensor. So is the diffusion tensor σ^{ij}, given by

$$1) \qquad E_t d\xi^i d\xi^j = \sigma^{ij}(\xi(t))dt ,$$

of a diffusion process ξ, where E_t denotes the conditional expectation given the configuration $\xi(t)$ at time t. We may set σ^{ij} proportional to m^{ij}, but dimensional considerations require the proportionality constant to have the dimensions of action. The *background field hypothesis* is that interaction with the background field causes the system to undergo a diffusion process with $\sigma^{ij} = \hbar m^{ij}$ satisfying a variational principle $\delta E \int L dt = 0$.

The paths of a diffusion process are nondifferentiable. This can be guessed from (1), which shows that $d\xi^i$ is of order $\sqrt{dt}$. This means that the kinematics of diffusion is very different from deterministic kinematics. What is meant by velocity and acceleration? How are tensors transported along paths? What is meant by an action integral involving the square of the velocity? How can one construct a diffusion process from an infinitesimal description of it? These questions are discussed in Chapter I. Next we take up stochastic quantization, basing it on a variational principle, and derive the Schrödinger equation. Then familiar topics from quantum mechanics – interference, momentum, bound states, statistics, spin – are discussed. Following this we investigate the physical interpretation of the theory, including problems of measurement and nonlocality. The notes conclude with a brief discussion of stochastic field theory.

Quantum

Fluctuations

Chapter I
KINEMATICS OF DIFFUSION

This chapter is long, and much of the material in it is well known. But some readers may be familiar with differential geometry but not measures on path space, or vice versa, so I have included a lot of elementary material for reference.

Most of the differential geometric material is physically relevant only to the discussion of spin, and it may be omitted by the reader who is willing to accept the result that one obtains the correct equation of motion even on a curved manifold.

The main exposition of this chapter begins in §5 where I attempt to construct a stochastic calculus with a minimum of technicalities. This is done at the cost of a loss of generality and elegance, and for a better approach the reader is referred to the French and Japanese schools of probability theory, e.g. [4], [32], [36], [45].

§1. Differentiable Manifolds

The first two sections of this chapter are a review of deterministic kinematics.

The configuration space of a mechanical system is a differentiable manifold M. If the system has n degrees of freedom, then M is an n-dimensional manifold. An n-dimensional manifold is specified as follows. It is a pair consisting of a set M and a collection $\mathfrak{U}$, called an *atlas*, of *charts*. A chart is a pair (U, Φ_U) consisting of a subset U of M, called a *coordinate neighborhood*, and a bijective mapping Φ_U from U to an open subset of $\mathbf{R}^n$. The components $q^1, \cdots, q^n$ of Φ_U are called *local coordinates*. If U' is also a coordinate neighborhood, then we have two systems of local coordinates, $q^1, \cdots, q^n$ and $q^{1'}, \cdots, q^{n'}$, on $U \cap U'$. (It is convenient to use different index sets $1, \cdots, n$ and $1', \cdots, n'$ for different local coordinates.) We require that the mapping $\Phi_{U'} \circ \Phi_U^{-1}$ be C^∞. Its matrix of partial derivatives is denoted by $\partial q^{i'}/\partial q^i$. We give M a topology by defining a subset G of M to be open in case each $\Phi_U(G \cap U)$ is open. We also require that M be a Hausdorff space; that is, for any two distinct points x_1 and x_2 in M there are disjoint open sets G_1 and G_2 containing them. Finally, we require that the coordinate neighborhoods cover M, and that M be connected.

For example, R^n is a manifold if we take the atlas consisting of a single chart: R^n and the identity mapping. If $n = 3N$, this is the configuration space of N spinless particles in 3-space.

Let M and N be differentiable manifolds. We say that $\Phi : M \to N$ is C^∞, or *smooth*, in case each $\Phi_V \circ \Phi \circ \Phi_U^{-1}$ is C^∞ (where U is a coordinate neighborhood in M and V is a coordinate neighborhood in N). If Φ is bijective and Φ^{-1} is also smooth, then Φ is called a *diffeomorphism* of M onto N, and M and N are *diffeomorphic* in case there exists such a Φ. Diffeomorphic manifolds may be regarded as being intrinsically the same.

The configuration of a mechanical system may change with time. Let ξ_1 and ξ_2 be two smooth paths in M starting at the same point x, so that $\xi_1(0) = \xi_2(0) = x$. Let $q^1, \cdots, q^n$ be local coordinates at x. We say that ξ_1 and ξ_2 are *tangent* to each other in case $\xi_1^i(t) - \xi_2^i(t) = o(t)$. (We use the notation $\xi^i(t) = q^i(\xi(t))$.) This happens if and only if $\dot{\xi}_1^i(0) = \dot{\xi}_2^i(0)$, where the dot denotes differentiation with respect to t. This condition is independent of the choice of local coordinates, for if $q^{i'}$ are also local coordinates at x, then

$$\dot{\xi}_1^{i'}(0) = \frac{\partial q^{i'}}{\partial q^i} \dot{\xi}_1^i(0) ,$$

and similarly for ξ_2. (The summation convention is used: any index occurring twice, once as an upper index and once as a lower index, is summed.) This allows us to give an intrinsic definition of a tangent vector: a *tangent vector* at x is an equivalence class of smooth paths starting at x and tangent to each other. In local coordinates, a tangent vector is given by an n-tuple of components v^i, and under a change of local coordinates they transform according to

$$v^{i'} = \frac{\partial q^{i'}}{\partial q^i} v^i .$$

The *velocity vector* of a smooth path ξ starting at x is its own equivalence class, so that in local coordinates it has components $\dot{\xi}^i(0)$.

The set of all tangent vectors at x forms an n-dimensional vector space, the *tangent space* T_xM. The set of all tangent vectors forms the *tangent bundle* TM. A point in TM is a pair (x,v) where $x \in M$ and $v \in T_xM$. The mapping $\pi : TM \to M$ given by $(x,v) \mapsto x$ is called the *projection* of TM onto M. The tangent bundle is a differentiable manifold in a natural way: the coordinate neighborhoods are the $\pi^{-1}U$, where U is a coordinate neighborhood in M, with local coordinates $q^1, \cdots, q^n$, $v^1, \cdots, v^n$. The tangent bundle is the *velocity phase space* of the mechanical system with configuration space M. A *state* of the system is a point in TM, and a *dynamical variable* is a function on $TM \times \mathbf{R}$, where $\mathbf{R}$ is the time axis.

A *vector field* on a manifold M is an assignment $x \mapsto X(x)$ of a tangent vector X(x) at x to each point x of M; that is, $x : M \to TM$ and $\pi X(x) = x$ for all x in M. A *scalar* is a function $f : M \to \mathbf{R}$. The smooth vector fields act on the smooth scalars as follows:

$$Xf(x) = X(x)^i \frac{\partial}{\partial q^i} f .$$

Since $\frac{\partial}{\partial q^{i'}} = \frac{\partial q^i}{\partial q^{i'}}\frac{\partial}{\partial q^i}$, $X(x)^{i'} = \frac{\partial q^{i'}}{\partial q^i} X(x)^i$, and $\frac{\partial q^{i'}}{\partial q^i}\frac{\partial q^j}{\partial q^{i'}} = \delta^j_i$, this is independent of the choice of local coordinates. (If the path ξ is in the equivalence class $X(x)$, then $Xf(x) = \frac{d}{dt} f(\xi(t))\Big|_{t=0}$.) In other words, vector fields may be identified with first order partial differential operators without zero-order part, acting on scalars.

A smooth vector field X on a manifold M gives rise to a unique local flow. For each point x in M there is a smooth path $t \mapsto \xi(t)$, defined at least for small values of $|t|$, with $\xi(0) = x$, such that the velocity vector of $\xi(t)$ is $X(\xi(t))$; in local coordinates, $\frac{d}{dt}\xi^i(t) = X^i(\xi(t))$. (In kinematics we are usually interested in flows on velocity phase space, but since TM is itself a manifold this case is included.) This is a familiar elementary theorem, but it is the fundamental fact about kinematics and the proof that the flow is smooth is difficult unless it is approached in the right way, so I will present it in full, to serve as a model for the more complicated case of diffusion processes.

The problem is a local one, so we work on R^n. On R^n, we may identify a vector field with a mapping $X: R^n \to R^n$. For simplicity of formulation, we modify X outside the neighborhood in which we are interested so that it has compact support (multiply it by a smooth function of compact support that is 1 on the neighborhood of interest).

THEOREM 1.1. *Let* $X: R^n \to R^n$ *be* C^∞ *with compact support. Then there is a unique* C^∞ *mapping* $\Phi: R \times R^n \to R^n$ *such that*

$$\frac{d}{dt}\Phi(t,x) = X(\Phi(t,x)), \quad \Phi(0,x) = x \tag{1.1}$$

for all t *in* R *and* x *in* R^n. *If we let* Φ^t *be* $x \mapsto \Phi(t,x)$, *then* $t \mapsto \Phi^t$ *is a one-parameter group of diffeomorphisms of* R^n:

$$\Phi^{t_1+t_2} = \Phi^{t_1} \circ \Phi^{t_2}. \tag{1.2}$$

Proof. Let κ be a Lipschitz constant for X; that is,

$$|X(x_1) - X(x_2)| \leq \kappa |x_1 - x_2|$$

for all x_1 and x_2 in R^n, where $|\ |$ is the Euclidean norm on R^n. First we prove uniqueness. Let $I = [0, 1/2\kappa]$. If $\xi : I \to R^n$, let

$$\|\xi\| = \sup_{s \in I} |\xi(s)| .$$

The initial value problem (1.1) is equivalent to

$$\Phi(t,x) = x + \int_0^t X(\Phi(s,x))ds . \tag{1.3}$$

Let Φ_1 and Φ_2 be solutions of (1.3). Then

$$\|\Phi_1(\cdot,x) - \Phi_2(\cdot,x)\| \leq \sup_{t \in I} \int_0^t \kappa |\Phi_1(s,x) - \Phi_2(s,x)| ds \leq \frac{1}{2} \|\Phi_1(\cdot,x) - \Phi_2(\cdot,x)\| ,$$

so that $\|\Phi_1(\cdot,x) - \Phi_2(\cdot,x)\| = 0$. This gives uniqueness for $t \in I$, and in the same way one gets uniqueness for $1/2\kappa \leq t \leq 1/\kappa$, etc., and for negative t. The group law (1.2) is an immediate consequence of uniqueness.

To prove existence, we use the method of successive approximations. Let

$$\Phi_0(t,x) = x$$

$$\Phi_1(t,x) = x + \int_0^t X(\Phi_0(s,x))ds ,$$

and, inductively,

$$\Phi_n(t,x) = x + \int_0^t X(\Phi_{n-1}(s,x))ds .$$

Then one has $\|\Phi_{n+1}(\cdot,x)-\Phi_n(\cdot,x)\| \le \frac{1}{2}\|\Phi_n(\cdot,x)-\Phi_{n-1}(\cdot,x)\|$, so the $\Phi_n(\cdot,x)$ are a Cauchy sequence. Define $\Phi(t,x) = \lim_{n\to\infty} \Phi_n(t,x)$ for $t \in I$. Then (1.3) holds on I, and by (1.2) we get a solution for all t. The mapping Φ_0 is C^∞, and by induction each Φ_n is C^∞. For each mult index α, we get a similar estimate on $\|D^\alpha\Phi_{n+1} - D^\alpha\Phi_n\|$, if we use the chain rule and replace I by a smaller interval (depending on α). There fore Φ is C^∞. ■

The proof shows more than was stated in the theorem. If $X: R^n \to R^n$ then we have existence and uniqueness merely if X satisfies a global Lipschitz condition. Any additional smoothness that X may have is shared by the flow. The theorem and proof are easily extended to include time dependent vector fields.

If X is a smooth vector field on a manifold M, use the theorem to construct a local flow in each coordinate neighborhood, and extend it as much as possible. Let $\dot{M}$ be the one point compactification of M (that is, $\dot{M} = M \cup \{\infty\}$ where $\infty \notin M$, and a neighborhood of ∞ is defined to be the complement of a compact subset of M). If the local flow $\Phi(t,x)$ has been defined for $0 \le t < b$ and it is not the case that $\lim_{t\to\infty} \Phi(t,x) = \infty$, then there is a compact set K contained in M such that $\Phi(t,x) \in K$ for a sequence of values of t tending to b. By compactness, there is at least one limit point x_0 in K of these points, but there is a local flow defined at x_0 and so $\Phi(t,x)$ can be defined beyond b. This is a sketch of the proof of the following theorem:

THEOREM 1.2. *Let* X *be a smooth vector field on the manifold* M. *Then there exist unique* G *and* Φ *such that* G *is an open subset of* $R \times M$; *for each* x *in* M, *the set of all* t *such that* $(t,x) \in G$ *is an open interval* (a_x, b_x) *containing* 0; $\Phi: G \to M$ *is smooth*; $\Phi(0,x) = x$ *for all* x *in* M; $\frac{d}{dt} q^i(\Phi(t,x)) = X^i(\Phi(t,x))$ *in local coordinates at* $\Phi(t,x)$, *for all* (t,x) *in* G; *if* $a_x \ne -\infty$, *then* $\lim_{t\to a_x} \Phi(t,x) = \infty$; *and if* $b_x \ne \infty$, *then* $\lim_{t\to b_x} \Phi(t,x) = \infty$.

Another popular method for constructing the local flow is the *polygonal approximation method.* Choose $dt > 0$. Start from x at time 0, and let the configuration move with constant velocity $X(x)$ for time dt. What I have just said is meaningless, because the notion of moving with constant velocity is undefined unless an affine connection (see §2) is specified; however, the present construction is insensitive to this and we may use any local coordinates at x to express the notion. The configuration arrives at $\xi(dt)$; then let it move with constant velocity $X(\xi(dt))$ for time dt, and so on. In the limit as $dt \to 0$ we obtain a path ξ with velocity $X(\xi(t))$. (It is easier to let dt be infinitesimal and take ξ to be the standard part of the polynomial approximation.)

A *cotangent vector* at x is a linear functional on T_xM; the set of all cotangent vectors at x is the dual space T_x^*M. The components a_i of a cotangent vector a transform according to

$$a_{i'} = \frac{\partial q^i}{\partial q^{i'}} a_i \, .$$

The *cotangent bundle* T_M^* is the set of all cotangent vectors; it is a differentiable manifold with projection $\pi : T^*M \to M$. A *cotangent vector field* or *1-form* a is a section of this bundle; that is, $a : M \to T^*M$ and $\pi a(x) = x$ for all x in M. If f is a smooth scalar, we define the 1-form df by giving its values on vector fields X by $\langle df, X \rangle = Xf$. Notice that if $X_1(x) = X_2(x)$ then $\langle df, X_1 \rangle(x) = \langle df, X_2 \rangle(x)$ —to verify this, we need only observe that the expression $\langle df, X \rangle$ is linear in X over the scalars, $\langle df, gX \rangle = g \langle df, X \rangle$ —so df is a well-defined 1-form. In local coordinates, $df_i = \frac{\partial}{\partial q_i} f$.

We let $T_{xr}^{\ s}M$ be $T_xM \otimes \cdots \otimes T_xM \otimes T_x^*M \cdots T_x^*M$ with r factors T_xM and s factors T_M^*. An element a of $T_{xr}^{\ s}M$ is called a *tensor* at x, of *contravariant rank* r and *covariant rank* s (or of *type* (r,s)). Its components transform according to

$$a^{i_1',\dots,i_r'}_{j_1',\dots,j_s'} = \frac{\partial q^{i_1'}}{\partial q^{i_1}} \cdots \frac{\partial q^{i_r'}}{\partial q^{i_r}} \frac{\partial q^{j_1}}{\partial q^{j_1'}} \cdots \frac{\partial q^{j_s}}{\partial q^{j_s'}} a^{i_1\dots i_r}_{j_1\dots j_s} .$$

The corresponding *tensor bundle* $T^s_r M$ is a differentiable manifold with projection $\pi : T^s_r M \to M$, and a *tensor field* is a section of this bundle. The set of smooth tensor fields of type (r,s) is denoted by $\widetilde{T}^s_r M$.

If X is a smooth vector field, we may differentiate a tensor field a along the paths of its local flow, obtaining the *Lie derivative* $\theta_X a$. On scalars f we have $\theta_X f = Xf$ and on vector fields Y we have $\theta_X Y = [X,Y]$, the Lie product. On scalars f we have $[X,Y]f = (XY-YX)f$, and in local coordinates

$$[X,Y]^i = X^j \frac{\partial}{\partial q^j} Y^i - Y^j \frac{\partial}{\partial q^j} X^i .$$

Once we know a derivation ϕ (such as the Lie derivative θ_X) on scalars and vector fields, its values on arbitrary tensor fields are determined by the product rule for differentiation. For a 1-form a the equation

$$\phi\langle a,Y\rangle = \langle \phi a,Y\rangle + \langle a,\phi Y\rangle \tag{1.4}$$

serves to define ϕa, and then we have

$$\phi(b\otimes c) = \phi b\otimes c + b\otimes\phi c$$

for arbitrary tensor fields b and c. A type of derivation that arises frequently is the one induced by a linear transformation on the tangent space. We extend it to all tensors by letting it be 0 on the scalars. Then the left-hand side of (1.4) is 0, so on cotangent vectors it is the negative transpose. If the matrix of the linear transformation is ϕ^i_j, I will denote the action of the induced derivation on a tensor a of type (r,s) by $\phi\colon a$. In local coordinates,

$$\phi\colon a^{i_1\dots i_r}_{j_1\dots j_s} = \sum_{\mu=1}^{r} \phi^{i_\mu}_{i} a^{i_1\dots i_{\mu-1} i\, i_{\mu+1}\dots i_r}_{j_1\dots j_s} - \sum_{\nu=1}^{s} \phi^{j}_{j_\nu} a^{i_1\dots i_r}_{j_1\dots j_{\nu-1} j\, j_{\nu+1}\dots j_s} . \tag{1.5}$$

§2. Affine Connections

In the previous section we discussed the velocity vector of a smooth path ξ in the configuration space M. Now let us discuss its acceleration.

There is a problem here. If we attempt to define the acceleration as

$$\lim_{dt\to 0} \frac{X(t+dt)-X(t)}{dt}, \tag{2.1}$$

where $X(t)$ is the velocity vector of ξ at time t, the attempt makes no sense: $X(t+dt)$ and $X(t)$ live in different vector spaces ($T_{\xi(t+dt)}M$ and $T_{\xi(t)}M$), and we cannot subtract them. If we use local coordinates q^i and consider the second derivative $\ddot{\xi}^i$, we find that under a change of local coordinates

$$\ddot{\xi}^{i'} = \frac{\partial q^{i'}}{\partial q^i}\ddot{\xi}^i + \frac{\partial^2 q^{i'}}{\partial q^j \partial q^k}\dot{\xi}^j\dot{\xi}^k .$$

Thus $\ddot{\xi}^i$ is not a vector; it may be 0 in one coordinate system but not in another.

Heuristically, what we need to make sense of (2.1), or more generally of

$$\lim_{dt\to 0} \frac{Y(\xi(t+dt))-Y(\xi(t))}{dt}$$

where Y is any vector field, is an identification τ of tangent spaces at neighboring (infinitely close) points in M. Then we could let $\nabla_X Y$, where $X = \dot{\xi}$, be

$$\lim_{dt\to 0} \frac{\tau Y(\xi(t+dt))-Y(\xi(t))}{dt}. \tag{2.2}$$

Then ∇ should have the following properties: ∇ is a mapping of $\tilde{T}M \times \tilde{T}M$ into $\tilde{T}M$ (where $\tilde{T}M = \tilde{T}^0_1 M$ is the smooth vector fields) such that

$$\nabla_X(Y+Z) = \nabla_X Y + \nabla_X Z, \tag{2.3}$$

(2.4) $$\nabla_{fX+gY} = f\nabla_X + g\nabla_Y ,$$

(2.5) $$\nabla_X(fY) = f\nabla_X Y + (Xf)Y .$$

A mapping ∇ with these properties is called an *affine connection.* An affine connection is an additional structure that must be imposed on a manifold M; there is no intrinsic affine connection on a differentiable manifold. (Later we will be concerned exclusively with Riemannian manifolds, for which there is a distinguished affine connection, the Riemannian connection.)

For the remainder of this section we consider a manifold M with a given affine connection ∇.

If f is a smooth scalar, we define $\nabla_X f$ to be Xf, and then we extend ∇_X to be a derivation on smooth tensor fields (see §1). Notice that, by (2.4), the action of ∇_X at x is determined by knowing $X(x)$; this is in contrast to the Lie derivative θ_X for which we need to know X in a neighborhood of x (or at least its first derivatives at x). If $a \in \widetilde{T}^s_r M$, we define the *covariant derivative* ∇a in $\widetilde{T}^{s+1}_r M$ by $\langle \nabla a, X\rangle = \nabla_X a$; this is linear in X over the scalars by (2.4), so ∇a is well defined. In local coordinates q^i, we denote $\nabla_{\frac{\partial}{\partial q^i}}$ by ∇_i and define the *Christoffel symbols* Γ^i_{jk} by

(2.6) $$\nabla_j \frac{\partial}{\partial q^k} = \Gamma^i_{jk} \frac{\partial}{\partial q^i} .$$

This is not a tensor; under a change of local coordinates we have

(2.7) $$\Gamma^{i'}_{j'k'} = \frac{\partial q^{i'}}{\partial q^i} \frac{\partial q^j}{\partial q^{j'}} \frac{\partial q^k}{\partial q^{k'}} \Gamma^i_{jk} + \frac{\partial^2 q^a}{\partial q^{j'} \partial q^{k'}} \frac{\partial q^{i'}}{\partial q^a} .$$

Then if $a \in \widetilde{T}^s_r M$, the formula for $\nabla_j a$ is

(2.8) $$\nabla_j a = \frac{\partial}{\partial q^j} a + \Gamma^{\cdot}_{j\cdot} a$$

(see (1.5)).

The *acceleration vector* of a smooth path ξ is defined to be $\nabla_X X$, where X is the velocity vector. (X is defined only on ξ. Extend it to be a smooth vector field $\tilde{X}$. Then $\nabla_X \tilde{X}$ is independent of the extension chosen, since X is tangent to ξ.) In local coordinates the acceleration vector a is given by

$$a^i = \ddot{\xi}^i + \Gamma^i_{jk}\dot{\xi}^j\dot{\xi}^k . \tag{2.9}$$

The *torsion* T of ∇ is defined by $T(X,Y) = \nabla_X Y - \nabla_Y X - [X,Y]$, and T is called *torsion-free* in case $T = 0$. In terms of the Christoffel symbols, ∇ is torsion-free if and only if $\Gamma^i_{jk} = \Gamma^i_{kj}$, as we see from (2.6) since $\left[\frac{\partial}{\partial q^j}, \frac{\partial}{\partial q^k}\right] = 0$. The torsion is clearly irrelevant to the acceleration (2.9), since $\dot{\xi}^j\dot{\xi}^k$ is symmetric in j and k.

From now on we assume that ∇ is torsion-free.

The *curvature* R of ∇ is defined by

$$R(X,Y) = \nabla_X\nabla_Y - \nabla_Y\nabla_X - \nabla_{[X,Y]} . \tag{2.10}$$

Thus $R\left(\frac{\partial}{\partial q^k}, \frac{\partial}{\partial q^\ell}\right)$ is the commutator of ∇_k and ∇_ℓ. The *curvature tensor*, also denoted by R, is given by

$$R(a,Z,X,Y) = \langle a, R(X,Y)Z \rangle$$

where a is a 1-form. To see that it is a tensor, we need only verify that it is linear over the scalars in each of its arguments, which is trivial to do. A simple computation starting from (2.6) shows that the components of R are given by

$$R^i{}_{jk\ell} = \frac{\partial}{\partial q^k}\Gamma^i_{\ell j} - \frac{\partial}{\partial q^\ell}\Gamma^i_{kj} + \Gamma^a_{\ell j}\Gamma^i_{ka} - \Gamma^a_{kj}\Gamma^i_{\ell a} . \tag{2.11}$$

We extend $R(X,Y)$ to be a derivation on the tensor algebra that is 0 on the scalars; see (1.5). Since $\left[\frac{\partial}{\partial q^k}, \frac{\partial}{\partial q^\ell}\right] = 0$,

(2.12) $$\nabla_k \nabla_\ell a - \nabla_\ell \nabla_k a = R^{\cdot}{}_{.k\ell} a \, .$$

This is all right as far as it goes. We have defined ∇_k to be $\nabla_{\frac{\partial}{\partial q^k}}$, and this maps $\widetilde{T}^s_r M$ into itself. But the components of $\nabla_k a$ are the components of the covariant derivative ∇a in $\widetilde{T}^{s+1}_r M$; the k also serves as a tensor index. Thus (2.12) is also open to a different interpretation, with ∇_k acting on the tensor index ℓ and vice versa, and the question arises as to whether it remains true for this different (and more useful) interpretation. The answer is yes, but only because our affine connection is torsion-free. Let us see why.

Recall that $\langle \nabla a, X \rangle = \nabla_X a$. Then $\langle \nabla\nabla a, X \otimes Y \rangle = \langle \nabla_X \nabla a, Y \rangle$. Since ∇_X is a derivation,

$$\nabla_X \langle \nabla a, Y \rangle = \langle \nabla_X \nabla a, Y \rangle + \langle \nabla a, \nabla_X Y \rangle \, ,$$

and $\langle \nabla a, Y \rangle = \nabla_Y a$ and $\langle \nabla a, \nabla_X Y \rangle = \nabla_{\nabla_X Y} a$, so

$$\langle \nabla_X \nabla a, Y \rangle = \nabla_X \nabla_Y a - \nabla_{\nabla_X Y} a \, .$$

Therefore

(2.13) $$\langle \nabla\nabla a, X \otimes Y \rangle - \langle \nabla\nabla a, Y \otimes X \rangle = (\nabla_X \nabla_Y - \nabla_Y \nabla_X - (\nabla_{\nabla_X Y} - \nabla_{\nabla_Y X}))\, a \, .$$

But since ∇ is torsion-free, $\nabla_X Y - \nabla_Y X = [X,Y]$; so by (2.10), (2.13) is

(2.14) $$\langle \nabla\nabla a, X \otimes Y \rangle - \langle \nabla\nabla a, Y \otimes X \rangle = R(X,Y)\, a \, .$$

A common abuse of notation is to write $(\nabla\nabla a)^{i_1 \cdots i_r}_{k\ell j_1 \cdots j_s}$ as $\nabla_k \nabla_\ell a^{i_1 \cdots i_r}_{j_1 \cdots j_s}$; with this convention, (2.14) in local coordinates $\left(\text{for } X = \frac{\partial}{\partial q^k} \text{ and } Y = \frac{\partial}{\partial q^\ell}\right)$ is written as (2.12). This is *Ricci's identity.*

Let ξ be a smooth path in M with velocity vector X and let Y_0 be in $T_{\xi(s)} M$. Then we may ask for a family $Y(t)$ in $T_{\xi(t)} M$ such that

(2.15) $$\nabla_{X(t)}Y(t) = 0, \quad Y(s) = Y_0 .$$

In local coordinates, (2.15) is

(2.16) $$\frac{dY^i}{dt} + \Gamma^i_{jk}X^jY^k = 0, \quad Y^i(s) = Y^i_0 .$$

This is a linear differential equation for Y, and its solution for arbitrary Y_0 gives a linear mapping $\tau_\xi(s,t) : T_{\xi(s)}M \to T_{\xi(t)}M$. This is *parallel translation.* There is an induced linear mapping, again denoted by $\tau_\xi(s,t)$, of $T^p_{\xi(s)r}$ into $T^p_{\xi(t)r}$.

Using this, we can make precise the heuristic notion used to motivate the definition of an affine connection: $\nabla_X Y$ is (2.2) if the previously undefined τ in it is taken to be $\tau_\xi(t+dt, t)$.

Let ξ be a smooth path. Each Y in $T_{\xi(t)}M$ is an equivalence class of mutually tangent smooth curves $r \mapsto \eta(r)$ with $\eta(0) = \xi(t)$. Let $y = \eta(dr)$; this is well defined up to $o(dr)$. That is, we may think of a tangent vector Y as being a neighboring point y. On an affinely connected manifold, parallel translation moves an infinitesimal neighborhood of $\xi(t)$ in such a way that the points in it all have the same velocity as $\dot\xi(t)$.

§3. Measures on Path Space

You all know probability theory in depth, thanks to the labors among you of the apostle of probability to Swiss physics. Nevertheless, let me review the basics.

A *probability space* is a measure space of total measure one — that is, a triple $(S, \mathcal{S}, \mu)$ where S is a set, $\mathcal{S}$ is a σ-algebra of subsets of S, and μ is a positive countably additive function on $\mathcal{S}$ such that $\mu(S) = 1$. A *random variable* f is a measurable function on S. If it is numerical valued, we define its *expectation* (or *mean*) by

$$Ef = \int f d\mu$$

and its *variance* by $E(f-Ef)^2$, provided the integrals exist. An *event* is an element of $\mathcal{S}$. A *stochastic process* is a function from some index set I to the random variables on a probability space.

One peculiarity of probabilists is that they take σ-algebras seriously, rather than regard them as a technical nuisance. Let $\mathcal{B}$ be a sub-σ-algebra of $\mathcal{S}$. Then $(S, \mathcal{B}, \mu\restriction\mathcal{B})$ is also a probability space. If $f \in L^1(S, \mathcal{S}, \mu)$, then $A \mapsto \int_A f d\mu$ for A in $\mathcal{B}$ is a finite measure on $(S, \mathcal{B})$ that is absolutely continuous with respect to μ, so by the Radon-Nikodym theorem there is a unique (up to equality a.e.) element f_0 of $L^1(S,\mathcal{B},\mu\restriction\mathcal{B})$ such that $\int_A f d\mu = \int_A f_0 d\mu$ for all A in $\mathcal{B}$. Then f_0 is called the *conditional expectation* of f with respect to $\mathcal{B}$, and it is denoted by $E\{f|\mathcal{B}\}$. We may think of $\mathcal{B}$ as being the events accessible to observation; then $E\{f|\mathcal{B}\}$ is the best prediction of f that we can make on the basis of available knowledge. We denote by $\tilde{\mathcal{B}}$ the set of $\mathcal{B}$-measurable random variables. Then $E\{f|\mathcal{B}\}$ is $\tilde{\mathcal{B}}$-linear: $E\{gf|\mathcal{B}\} = gE\{f|\mathcal{B}\}$ for g in $\tilde{\mathcal{B}}$, provided the integrals exist.

Suppose we want to study a stochastic process indexed by a set I and taking values in a locally compact Hausdorff space M. Form the one-point compactification $\dot{M} = M \cup \{\infty\}$. This is a compact Hausdorff space. Now form the Cartesian product Ω of $\dot{M}$ indexed by I:

$$\Omega = \prod_I \dot{M} .$$

An element of Ω is a completely arbitrary function $\omega: I \to \dot{M}$. By the Tychonov theorem, Ω is a compact Hausdorff space in the product topology (defined by the requirement that a net $\alpha \mapsto \omega_\alpha$ converges in Ω to ω if and only if $\omega_\alpha(t)$ converges in $\dot{M}$ to $\omega(t)$ for each t in I). We call Ω *path space*.

Consider an M-valued stochastic process ξ indexed by I and defined over a probability space $(S, \mathcal{S}, \mu)$. This induces a probability measure on path space Ω in a natural way, as follows. Let $C(\Omega)$ be the algebra of all continuous real-valued functions f on Ω. By the Riesz-Markov theorem, there is a one-to-one correspondence between positive linear functionals L on $C(\Omega)$ such that $L(1) = 1$ and regular probability measures Pr on Ω, the correspondence being such that $L(f) = \int_\Omega f(\omega)\, d\Pr(\omega)$ for all f in $C(\Omega)$. (A measure is called *regular* in case the σ-algebra on which it is defined is the σ-algebra $\mathcal{B}$ of Borel sets – the smallest σ-algebra containing all compact sets – and is such that

$$\Pr(B) = \sup\{\Pr(K) : K \subseteq B \text{ and } K \text{ is compact}\} \tag{3.1}$$

for all Borel sets B.) Let $C_{fin}(\Omega)$ be the subalgebra of $C(\Omega)$ consisting of those f depending on only finitely many indices (so that there is a finite subset I_0 of I such that for all ω_1 and ω_2 in Ω, if $\omega_1 \restriction I_0 = \omega_2 \restriction I_0$ then $f(\omega_1) = f(\omega_2)$). Then $C_{fin}(\Omega)$ contains 1 and separates points, so by the Stone-Weierstrass theorem it is dense in $C(\Omega)$ in the supremum norm $\|f\| = \sup\{|f(\omega)| : \omega \in \Omega\}$. Hence any positive linear functional L on $C_{fin}(\Omega)$ such that $L(1) = 1$ has a unique extension to $C(\omega)$ with the same properties. Now any f in $C_{fin}(\Omega)$ is of the form

$$f(\omega) = F(\omega(t_1), \cdots, \omega(t_n)) \tag{3.2}$$

for some finite subset $I_0 = \{t_1, \cdots, t_n\}$ of I and F in $C(\dot{M}^n)$. Define

$$L(f) = \int_S F(\xi(t_1), \cdots, \xi(t_n))\, d\mu$$

and let Pr be the corresponding regular probability measure on Ω. Then for each I_0 and F as above,

$$\int_S F(\xi(t_1), \cdots, \xi(t_n))\, d\mu = \int_\Omega F(\omega(t_1), \cdots, \omega(t_n))\, d\Pr(\omega) .$$

Path space Ω has the advantage of greater intuitive appeal and greater technical simplicity in the handling of events whose definition involves uncountably many t, by means of the following simple technical lemma:

THEOREM 3.1. *Let* Pr *be a regular probability measure, let* $\mathcal{G}$ *be a family of open sets that is closed under finite unions, and let* U *be the union of all sets in* $\mathcal{G}$. *Then* $\Pr(U) = \sup\{\Pr(G) : G \in \mathcal{G}\}$.

Proof. Let $\varepsilon > 0$. Since Pr is regular, there is a compact set K contained in U such that $\Pr(U) - \Pr(K) \leq \varepsilon$. By compactness, there are a finite number of sets $G_1, \cdots, G_n$ in $\mathcal{G}$ covering K. Let G be their union, so that $G \in \mathcal{G}$ and $K \subseteq G \subseteq U$. Then $\Pr(U) - \Pr(G) \leq \varepsilon$. Since ε is arbitrary, the conclusion holds. ■

The point of this theorem is that the family $\mathcal{G}$ may be uncountable. The openness is then essential: consider the family of finite subsets of the unit interval – their union has Lebesgue measure 1 but each set in the family has measure 0.

The use of regular probability measures on path space is due to Kakutani, and is developed in [46].

§4. Martingales

Consider a real-valued stochastic process ξ indexed by a subset I of $\mathbf{R}$, defined on a probability space $(S, \mathcal{S}, \mu)$, and such that each $\xi(t)$ is in L^1 (we will always tacitly assume that our time parameters lie in I). Let $\mathcal{P} : t \to \mathcal{P}_t$ be an increasing family of sub-σ-algebra of $\mathcal{S}$ (then $\mathcal{P}$ is called a *filtration*) such that each $\xi(t)$ is in $\tilde{\mathcal{P}}_t$ (then ξ is said to be *adapted* to $\mathcal{P}$). For example, $\mathcal{P}_t$ may be the σ-algebra generated by the $\xi(s)$ with $s \leq t$ (and it must contain this σ-algebra). We denote $E\{\cdot \mid \mathcal{P}_t\}$ by $\overline{E}_t$. We say that ξ is a *martingale* (with respect to $\mathcal{P}$) in case $\overline{E}_s \xi(t) = \xi(s)$ for $s \leq t$, a *supermartingale* in case $\overline{E}_s \xi(t) \leq \xi(s)$ for $s \leq t$, and a *submartingale* in case $\overline{E}_s \xi(t) \geq \xi(s)$ for $s \leq t$.

Suppose that I is a finite set, with first element a and last element b. We let $I' = I \setminus \{b\}$, for all t in I′ we let t+dt be its successor in I, and we define $d\xi(t) = \xi(t+dt) - \xi(t)$. We define $\bar{D}\xi$, $d\hat{\xi}$, and σ_ξ^2 by

$$\bar{D}\xi(t)dt = \bar{E}_t d\xi(t) , \tag{4.1}$$

$$d\xi(t) = \bar{D}\xi(t)dt + d\hat{\xi}(t) , \tag{4.2}$$

$$\sigma_\xi^2(t)dt = \bar{E}_t d\hat{\xi}(t)^2 . \tag{4.3}$$

Thus $\bar{D}\xi(t)dt$ and $\sigma_\xi^2(t)dt$ are the conditional mean and variance of $d\xi(t)$. Notice that $\bar{D}\xi$ and σ_ξ^2 are adapted to $\mathcal{P} \restriction I'$, whereas $d\xi$ and $d\hat{\xi}$ in general are not.

Notice that $\bar{D}\xi = 0$ if ξ is a martingale, $\bar{D}\xi \le 0$ if ξ is a supermartingale, and $\bar{D}\xi \ge 0$ if ξ is a submartingale. (Of course, for a general process $\bar{D}\xi$ need not have a constant sign either on I′ or on the probability space.) We will show that these conditions are sufficient as well as necessary. We have

$$\xi(t) = \xi(s) + \sum_{s \le r < t} \bar{D}\xi(r)dr + \sum_{s \le r < t} d\hat{\xi}(r) , \quad s \le t . \tag{4.4}$$

Now $\bar{E}_t d\hat{\xi}(t) = 0$, and since $\mathcal{P}$ is increasing we have $\bar{E}_s \bar{E}_t = \bar{E}_s$ for $s \le t$, so that

$$\bar{E}_s \xi(t) = \xi(s) + \bar{E}_s \sum_{s \le r < t} \bar{D}\xi(r)dr , \quad s \le t . \tag{4.5}$$

Therefore ξ is a martingale if and only if $\bar{D}\xi = 0$, etc.

We call $\bar{D}\xi$ the *trend* of ξ, and if $d\hat{\xi} = 0$ we say that ξ is a *trend process*. Thus ξ is a trend process if and only if $\sigma_\xi^2 = 0$ or equivalently $d\xi$ is adapted to $\mathcal{P} \restriction I'$. We let

$$\tilde{\xi}(t) = \xi(a) + \sum_{r < t} \bar{D}\xi(r)dr . \tag{4.6}$$

Then $d\tilde{\xi}(t) = \bar{D}\xi(t)dt \in \tilde{\mathcal{P}}_t$, so that $\tilde{\xi}$ is a trend process, called the *trend process associated with* ξ. We let

$$\hat{\xi}(t) = \sum_{r<t} d\hat{\xi}(r) , \tag{4.7}$$

so that $\hat{\xi}(a) = 0$. Thus $\hat{\xi}$ is adapted to $\mathcal{P}$ and its increments, as the notation requires, are the $d\hat{\xi}(t)$. Since $\bar{D}\hat{\xi} = 0$, the process $\hat{\xi}$ is a martingale. We call it the *martingale associated with* ξ. Notice that if ξ is already a martingale, then $\hat{\xi}$ differs trivially from it: $\hat{\xi}(t) = \xi(t) - \xi(a)$. We have the decomposition $\xi = \tilde{\xi} + \hat{\xi}$ of an arbitrary L^1 process ξ into a trend process $\tilde{\xi}$ and a martingale $\hat{\xi}$.

If we have uniform bounds on $\|\bar{D}\xi(t)\|_1$ and $\|\sigma^2_{\xi}(t)\|_1$, then $d\xi(t)$ will be of order dt but $d\hat{\xi}(t)$ will be much larger, of order $\sqrt{dt}$. The interesting local fluctuations of a process are in the associated martingale.

Let ξ and η be adapted to $\mathcal{P}$, and define ζ by

$$\zeta(t) = \sum_{r<t} \eta(r)d\xi(r) ; \tag{4.8}$$

then ζ is also adapted to $\mathcal{P}$. We have $\zeta(a) = 0$ and $d\zeta(t) = \eta(t)d\xi(t)$, so that (if $\zeta(t) \in L^1$)

$$\bar{D}\zeta(t) = \eta(t)\bar{D}\xi(t) . \tag{4.9}$$

Thus if ξ is a martingale, so is ζ; if $\eta \geq 0$, then if ξ is a super- or submartingale, so is ζ.

The key to proving theorems about martingale and super- or sub-martingales ξ is to think of $\xi(t)$ as being the price at time t of a share of a stock on the stock market. A martingale is a steady market, and a supermartingale is a declining market. Then η is an investment strategy: at each time t, $\eta(t)$ is the number of shares an investor holds. The requirement that η be adapted to $\mathcal{P}$ is the requirement that the investment strategy be legal: no inside information about the future behavior is allowed. The process ζ given by (4.8) is the investor's "earnings."

For example, suppose an investor buys one share at time a and keeps it until the price increases by more than λ, where $\lambda > 0$, and as soon as this happens the investor sells the share. That is,

$$\eta(t) = \begin{cases} 1, \xi(s) - \xi(a) \le \lambda & \text{for all } s \le t, \\ 0, & \text{otherwise.} \end{cases}$$

Let Λ be the event that the investor's strategy is successful,

$$\Lambda = \{\max(\xi(t) - \xi(a)) > \lambda\} .$$

(This is a subset of the underlying probability space S, and it is the custom to suppress the variable ranging over S.) The investor's earnings are $> \lambda$ if successful, and are $\xi(b) - \xi(a)$ otherwise. That is,

$$\zeta(b) \ge \lambda \chi_\Lambda + (\xi(b) - \xi(a)) \chi_{\Lambda^c} , \tag{4.10}$$

where χ denotes the indicator function and ^{c} denotes the complement. Now suppose that, unfortunately for the investor, ξ is a supermartingale. Since $\eta \ge 0$, ζ is also a supermartingale, and since $\zeta(a) = 0$ we have $E\zeta(b) \le 0$. By (4.10) this implies that $\Pr\Lambda \le \|\xi(b) - \xi(a)\|_1/\lambda$. We have proved the following theorem:

THEOREM 4.1. *Let ξ be a supermartingale indexed by a finite subset I of R, and let $\lambda > 0$. Then*

$$\Pr\{\max(\xi(t) - (a)) > \lambda\} \le \frac{1}{\lambda} \|\xi(b) - \xi(a)\|_1 .$$

We may remove the restriction that I be finite. In the following theorem, $\Pr$ is the regular probability measure on path space (as in the preceding section).

THEOREM 4.2. *Let* ξ *be a supermartingale and let* $\lambda > 0$. *Then*

$$\Pr\{\sup_{a \le t \le b} (\xi(t) - \xi(a)) > \lambda\} \le \frac{1}{\lambda} \|\xi(b) - \xi(a)\|_1 . \tag{4.11}$$

Proof. The events $\{\omega(t) - \omega(a) > \lambda\}$ are open, so (4.11) holds by Theorems 4.1 and 3.1. ■

Theorems 4.1 and 4.2 remain true if "supermartingale" is replaced by "submartingale." To see this, use the dual investment strategy: replace η by $1-\eta$. Also, if f is a convex function (e.g. $f(\xi) = (\xi - c)^2$) and ξ is a martingale, then $f \circ \xi$ is a submartingale. Therefore we have the following result:

THEOREM 4.3. *Let* ξ *be a martingale and let* $\lambda > 0$. *Then*

$$\Pr\{\sup_{a \le t \le b} |\omega(t) - \omega(a)| > \lambda\} \le \frac{1}{\lambda^2} \|\xi(b) - \xi(a)\|_2^2 .$$

The terminology "supermartingale" for a process with a negative trend may seem perverse, but there is a good reason for it. If f is a superharmonic function on $\mathbf{R}^n$ and w is the Wiener process (see §11), then $f \circ w$ is a supermartingale.

Each investment strategy leads to a martingale inequality. For example, the "buy low – sell high" strategy leads to Doob's upcrossing inequality and so to the martingale convergence theorem, but we shan't pursue this here; see [18] or [61] for a systematic study containing the results of this section and many others.

§5. Diffusion

Let M be a manifold, I an interval, and ξ an M-valued stochastic process indexed by I. We let the *past* $\mathcal{P}_t$, the *future* $\mathcal{F}_t$, and the *present* $\mathcal{N}_t$ be the σ-algebras generated by the $\xi(s)$ with $s \le t$, $s \ge t$, and $s = t$ respectively. We will abbreviate $E\{\cdot \mid \mathcal{N}_t\}$ by E_t.

We use dt as a strictly positive real variable, and for any function f of t we define the *forward differential* $df(t) = f(t+dt)-f(t)$. Let $\mathcal{B}_t$ be the set of all uniformly bounded real stochastic processes η indexed by some interval $[0, \varepsilon]$, and defined over the same probability space as ξ, such that $\|E_t\eta(dt)\|_\infty = O(dt)$ and $\|E_t\eta(dt)^2\|_\infty = O(dt)$. Then $\mathcal{B}_t$ is an algebra. Let $\mathcal{O}_t$ be the subset of all ζ in $\mathcal{B}_t$ such that $\|E_t\zeta(dt)\|_\infty = o(dt)$ and $\|E_t\zeta(dt)^2\|_\infty = o(dt)$. Then $\mathcal{O}_t$ is an ideal in the algebra $\mathcal{B}_t$, by the Schwarz inequality. We will denote congruence modulo this ideal by $\equiv$. This notion plays the same role in stochastic kinematics as tangency plays in deterministic kinematics.

We make the convention that local coordinates q^i are globally defined and have compact support (though of course they need not be a coordinate system outside their coordinate neighborhood).

We say that ξ is a *smooth diffusion* in case whenever the q^i are local coordinates for a coordinate neighborhood U, there exists smooth functions β^i and σ^{ij}, with σ^{ij} of strictly positive type on U, such that

$$E_t d\xi^i(t) \equiv \beta^i(\xi(t), t)\,dt\,, \tag{5.1}$$

$$d\xi^i(t)\,d\xi^j(t) \equiv \sigma^{ij}(\xi(t))\,dt\,, \tag{5.2}$$

and the same is true of the time-reversed process $\check{\xi}(t) = \xi(-t)$ indexed by $\check{I} = \{t : -t \in I\}$.

Notice that there is no conditional expectation E_t on the left-hand side of (5.2). This is the crucial assumption that rules out processes with jump discontinuities, such as the Poisson process. Our results can easily be generalized to the case that σ^{ij} is time dependent. There are many interesting problems involving processes where σ^{ij} is degenerate, but I will not discuss such processes. Notice that ξ is not assumed to be a Markov process.

THEOREM 5.1. *Let ξ be a smooth diffusion on* M *and let* f *be a smooth scalar with compact support on* $M \times I$. *If the* q^i *are local* *coordinates, then*

$$(5.3) \qquad d\xi^i d\xi^j d\xi^k \equiv 0\,,$$

$$(5.4) \quad df(\xi(t),t) \equiv \frac{\partial f}{\partial q^i}(\xi(t),t)\,d\xi^i + \frac{1}{2}\,\frac{\partial^2 f}{\partial q^i \partial q^j}(\xi(t),t)\,\sigma^{ij}\,dt + \frac{\partial}{\partial t} f(\xi(t),t)\,dt\,,$$

$$(5.5) \qquad E_t df(\xi,t),t) \equiv \left(\frac{1}{2}\,\sigma^{ij}\,\frac{\partial^2}{\partial q^i \partial q^j} + \beta^i\,\frac{\partial}{\partial q^i} + \frac{\partial}{\partial t}\right) f(\xi(t),t)\,dt\,.$$

If the $q^{i'}$ *are also local coordinates, then (on the intersection of the two coordinate neighborhoods)*

$$(5.6) \qquad d\xi^{i'} \equiv \frac{\partial q^{i'}}{\partial q^i}\,d\xi^i + \frac{1}{2}\,\frac{\partial^2 q^{i'}}{\partial q^i \partial q^j}\,\sigma^{ij}\,dt\,,$$

$$(5.7) \qquad \beta^{i'} = \frac{\partial q^{i'}}{\partial q^i}\,\beta^i + \frac{1}{2}\,\sigma^{ij}\,\frac{\partial^2 q^{i'}}{\partial q^i \partial q^j}\,,$$

$$(5.8) \qquad \sigma^{i'j'} = \frac{\partial q^{i'}}{\partial q^i}\,\frac{\partial q^{j'}}{\partial q^j}\,\sigma^{ij}\,.$$

Proof. We have $d\xi^i d\xi^j d\xi^k \equiv \sigma^{ij} dt\, d\xi^k$, so that $E_t d\xi^i d\xi^j d\xi^k \equiv \sigma^{ij} dt\, \beta^k dt = o(dt)$ and $E_t (d\xi^i d\xi^j d\xi^k)^2 \equiv E_t((\sigma^{ij})^2 dt^2 \sigma^{kk} dt) = o(dt)$. Thus (5.3) holds. The other results hold by this and Taylor's formula. ∎

We call

$$(5.9) \qquad \frac{1}{2}\,\sigma^{ij}\,\frac{\partial^2}{\partial q^i \partial q^j} + \beta^i\,\frac{\partial}{\partial q^i}$$

the *forward diffusion operator*. By (5.7) and (5.8) it is independent of the choice of local coordinates. Notice that $d\xi^i$ and β^i do not transform like vectors, but that σ^{ij} is a contravariant tensor. Since we have assumed it to be nondegenerate, it has an inverse σ_{ij} (such that $\sigma_{ij}\sigma^{jk} = \delta^k_i$). Then σ_{ij} is a covariant symmetric tensor of strictly positive type; in short, a *Riemannian metric*. It is intrinsically associated with the diffusion and gives the scale of the local fluctuations.

The Riemannian metric σ_{ij} gives T_xM the structure of a Euclidean space – an n-dimensional real Hilbert space. We may choose an orthonormal basis for T_xM, and by a linear change of local coordinates at x we obtain local coordinates q^i such that $\sigma_{ij}(x) = \delta_{ij}$. In general it is not possible to have $\sigma_{ij} = \delta_{ij}$ in a neighborhood of x, but the fundamental theorem of local Riemannian geometry shows that we can always choose local coordinates q^i such that $\sigma_{ij} = \delta_{ij} + 0(|q|^2)$ (where for simplicity of notation we assume that $q^i(x) = 0$), so that $\partial\sigma_{ij}/\partial q^k = 0$ at x. Such local coordinates are called *normal coordinates* (NC), and they are unique up to an orthogonal transformation and terms that are $0(|q|^3)$.

THEOREM 5.2. *There exist normal coordinates* q^i *at* x. *If the* $q^{i'}$ *are also normal coordinates at* x, *then there is an orthogonal matrix* A *such that* $q^{i'} = A^{i'}_i q^i + 0(|q|^3)$.

Proof. Let q^i be local coordinates such that $q^i(x) = 0$ and $\sigma_{ij}(x) = \delta_{ij}$. Let $\alpha_{ija} = \partial\sigma_{ij}/\partial q^a$ at x, so that

$$\sigma_{ij} = \delta_{ij} + \alpha_{ija} q^a + 0(|q|^2) . \tag{5.10}$$

Let $\bar{q}^k = q^k + \beta^k_{ab} q^a q^b$. We want to pick β^k_{ab} so that the $\bar{q}^k$ are normal coordinates. We have

$$q^i = \bar{q}^i - \beta^i_{ab} \bar{q}^a \bar{q}^b + 0(|\bar{q}|^3) ,$$

so that

$$\frac{\partial q^i}{\partial \bar{q}^k} = \delta^i_k - 2\, \beta^i_{ka} \bar{q}^a + 0(|\bar{q}|^2) .$$

Observe that we may replace q by $\bar{q}$ in (5.10). Then

$$(5.11)\qquad \bar{\sigma}_{k\ell} = \frac{\partial q^i}{\partial \bar{q}^k}\frac{\partial q^j}{\partial \bar{q}^\ell}\sigma_{ij}$$

$$= (\delta^i_k - 2\beta^i_{ka}\bar{q}^a)(\delta^j_\ell - 2\beta^i_{\ell a}\bar{q}^a)(\delta_{ij} + \alpha_{ija}\bar{q}^a) + O(|\bar{q}|^2)$$

$$= \delta_{k\ell} + (-2\beta^\ell_{ka} - 2\beta^k_{\ell a} + \alpha_{k\ell a})\bar{q}^a + O(|\bar{q}|^2)\,.$$

We choose $\beta^\ell_{ka} = \frac{1}{4}\alpha_{k\ell a}$. Since $\alpha_{k\ell a} = \alpha_{\ell ka}$, we have $\bar{\sigma}_{k\ell} = \delta_{k\ell} + O(|\bar{q}|^2)$, and the $\bar{q}^k$ are normal coordinates.

Now suppose that the q^k were normal coordinates to begin with, so that $\alpha_{k\ell a} = 0$, and let the $\bar{q}^k$ also be normal coordinates. After an orthogonal change of variables we can assume that $q^k = \bar{q}^k + \beta^k_{ab}q^a q^b + O(|q|^3)$. Since $q^a q^b = q^b q^a$, we can assume that $\beta^k_{ab} = \beta^k_{ba}$. We again have (5.11), with $\alpha_{k\ell a} = 0$ and the coefficient of $\bar{q}^a$ equal to 0, so that $\beta^\ell_{ka} = -\beta^k_{\ell a}$. Therefore

$$\beta^\ell_{ka} = -\beta^k_{\ell a} = -\beta^k_{a\ell} = \beta^a_{k\ell} = \beta^a_{\ell k} = -\beta^\ell_{ak} = -\beta^\ell_{ka}\,,$$

so that $\beta^\ell_{ka} = 0$. ■

This theorem may be used to carry over to a Riemannian manifold concepts, involving at most one derivative of tensors, that are well defined on Euclidean space. This may be called the *transfer principle* of local Riemannian geometry; we will use it on a number of occasions. On Euclidean space, the covariant derivative of any tensor field a is given simply by $\nabla_i a = \partial a/\partial q^i$.

On a Riemannian manifold, we define

$$\nabla_i a = \frac{\partial}{\partial q^i}a \quad (\mathrm{NC})\,.$$

This defines the *Riemannian connection* ∇. The Riemannian connection is clearly torsion-free and $\nabla_i \sigma_{jk} = 0$. If one permutes this cyclically in i, j, k and subtracts one equation from the sum of the other two, one finds

$$\Gamma^k_{ij} = \frac{1}{2}\,\sigma^{ka}\left(\frac{\partial}{\partial q^i}\,\sigma_{ij} + \frac{\partial}{\partial q^j}\,\sigma_{ai} - \frac{\partial}{\partial q^a}\,\sigma_{ij}\right),$$

but we will not need this formula.

The *curvature* of a Riemannian manifold is the curvature of its Riemannian connection.

If a_i is a 1-form, we define the vector field a^i to be $\sigma^{ij}a_j$, and vice versa. In general, we use the metric tensor σ_{ij} and its inverse σ^{ij} to lower and raise indices. For example, the *Ricci tensor* is defined to be $R^i_j = R^i{}_a{}^a{}_j = \sigma^{ab}R^i{}_{abj}$ and the *scalar curvature* is $\bar{R} = R^i_i$.

The *volume element* $d_M x$ is defined by $d_M x = dx^1 \cdots dx^n$ (NC). Then we have $(X^i\nabla_i)^* f = -\nabla_i(X^i f)$, where * denotes the formal adjoint with respect to the volume element, since this clearly holds in normal coordinates.

For a smooth scalar f, we define the *Laplacian* Δf by $\Delta f = \delta^{ij}\dfrac{\partial^2 f}{\partial q^i \partial q^j}$ (NC). Since $\partial f/\partial q^j$ is a tensor (the 1-form df_j), this involves only one derivative of a tensor, and $\Delta f = \nabla^i\nabla_i f$ since this equation holds in normal coordinates. By contrast, the Laplacian of a general tensor field a cannot be defined to be $\delta^{ij}\dfrac{\partial^2 a}{\partial q^i \partial q^j}$ (NC) because the expression depends on the choice of normal coordinates. There are several inequivalent notions of the Laplacian of a tensor field, and we will see in §10 that one of them arises naturally in diffusion theory.

After this brief excursion into local Riemannian geometry, let us return to the diffusion process that gave rise to the Riemannian metric.

We observed that $d\xi^i$ does not transform like a vector. To remedy this, we define the *forward vector differential* $\tilde{d}\xi^i$ by setting $\tilde{d}\xi^i \equiv d\xi^i$ (NC) and insisting that it transform like a vector. Then

$$\tilde{d}\xi^{i'} \equiv \frac{\partial q^{i'}}{\partial q^i}\,d\xi^i .$$

To compute this, use (5.6), reversing the roles of the two coordinate systems. Then

$$\tilde{d}\xi^{i'} \equiv \frac{\partial q^{i'}}{\partial q^{i}}\left(\frac{\partial q^{i}}{\partial q^{a'}} d\xi^{a'} + \frac{1}{2}\,\frac{\partial^2 q^{i}}{\partial q^{a'}\partial q^{j'}}\,\sigma^{a'j'}dt\right)$$

$$\equiv d\xi^{i'} + \frac{1}{2}\,\Gamma^{i'}_{a'j'}\sigma^{a'j'}dt$$

by (2.7), since $\Gamma^{i}_{jk} = 0$. Thus in general coordinates we have

$$\tilde{d}\xi^{i} \equiv d\xi^{i} + \frac{1}{2}\,\Gamma^{i}_{jk}\,\sigma^{jk}dt\,. \tag{5.12}$$

Observe that $\tilde{d}\xi^{j}\tilde{d}\xi^{k} \equiv \sigma^{jk}dt$, so we may rewrite (5.12) as

$$d\xi^{i} \equiv \tilde{d}\xi^{i} - \frac{1}{2}\,\Gamma^{i}_{jk}\tilde{d}\xi^{j}\tilde{d}\xi^{k}\,. \tag{5.13}$$

Now the probabilistic meaning of $\tilde{d}\xi^{i}$ is clear from (5.13) and (2.9): if a configuration starts at $\xi(t)$ with velocity $\tilde{d}\xi^{i}/dt$ and moves freely for time dt, it will arrive at $\xi(t+dt)$ to within $o(dt)$.

Similarly, we define the *forward drift* $b^{i}(x,t)$ by setting $b^{i}(x,t) = \beta^{i}(x,t)$ (NC) and insisting that it transform like a vector. Then in general coordinates we have

$$b^{i}(x,t) = \beta^{i}(x,t) + \frac{1}{2}\,\Gamma^{i}\,, \tag{5.14}$$

where we have set $\Gamma^{i} = \Gamma^{i}_{jk}\,\sigma^{jk} = \Gamma^{ij}_{j}$. Observe that

$$E_t\tilde{d}\xi^{i} \equiv b^{i}(\xi(t),t)\,dt\,. \tag{5.15}$$

We may rewrite the forward diffusion operator (5.9) as $\frac{1}{2}\,\Delta + b^{i}\nabla_{i}$, (5.5) as

$$E_t df(\xi(t),t) \equiv \left(\frac{1}{2}\,\Delta + b^{i}\nabla_{i} + \frac{\partial}{\partial t}\right) f(\xi(t),t)\,dt\,,$$

and (5.4) as

$$\text{(5.16)}\qquad df(\xi(t),t) \equiv \frac{\partial f}{\partial q^i}(\xi(t),t)\,d\xi^i + \frac{1}{2}\Delta f(\xi(t),t)\,dt + \frac{\partial f}{\partial t}(\xi(t),t)\,dt\,.$$

Let ρ be the measure on $M \times I$ such that $\int_{M\times I} f\,d\rho = \int_I Ef(\xi(t),t)\,dt$. Let I^0 be the interior of I, and let f be in $C_0^\infty(M \times I^0)$, where C_0^∞ denotes the set of smooth functions with compact support. By (5.16),

$$0 = \int_I \frac{d}{dt} Ef(\xi(t),t)\,dt = \int_I E\left(\frac{1}{2}\Delta + b^i\nabla_i + \frac{\partial}{\partial t}\right) f(\xi(t),t)\,dt$$

$$= \int_{M\times I} \left(\frac{1}{2}\Delta + b^i\nabla_i + \frac{\partial}{\partial t}\right) f\,d\rho\,.$$

That is, ρ is a weak solution of

$$\frac{\partial\rho}{\partial t} = \frac{1}{2}\Delta\rho - \nabla_i(b^i\rho)\,.$$

But this equation — the *forward Fokker-Planck equation* — is parabolic, so that ρ is a smooth solution of it on $M \times I^0$ and is strictly positive. The function $\rho(x,t)$ is called the *probability density*. We denote $\rho(x,t)\,d_M x$ by $\rho(dx,t)$.

The definition of a smooth diffusion requires that the time reversed process $\check{\xi}$ also be a smooth diffusion. For any function $f(t)$, we define $\check{f}(t) = f(-t)$. We retain the convention that $dt > 0$, and define the *backward differential* $d_* f(t) = f(t) - f(t-dt)$, so that $d_* f(t) = df(t-dt) = -d\check{f}(-t)$. Then there are β_*^i and σ_*^{ij} such that

$$E_t d_*\xi^i(t) = \beta_*^i(\xi(t),t)\,dt\,,$$

$$d_*\xi^i(t)\,d_*\xi^j(t) = \sigma_*^{ij}(\xi(t))\,dt\,.$$

For a real-valued stochastic process F we define the *forward stochastic derivative* D and the *backward stochastic derivative* D_* by

$$(5.18)\qquad \begin{cases} DF(t) = \lim\limits_{dt\to 0+} E_t \dfrac{F(t+dt)-F(t)}{dt}\,, \\[2ex] D_*F(t) = \lim\limits_{dt\to 0+} E_t \dfrac{F(t)-F(t-dt)}{dt}\,, \end{cases}$$

whenever the conditional expectations and limits exist. Notice that E_t is the conditional expectation with respect to the present. Let f and g be in $C_0^\infty(M)$, and let $F(t) = f(\xi(t))$ and $G(t) = g(\xi(t))$. I claim that

$$(5.19)\qquad \int_a^b E(DF(t))G(t)\,dt = -\int_a^b EF(t)D_*G(t)\,dt + EF(b)G(b) - EF(a)G(a)\,.$$

To prove this, divide $[a,b]$ into ν equal parts: $t_j = a + j(b-a)/\nu$ for $j = 0, \cdots, \nu$. Then

$$EF(b)G(b) - EF(a)G(a) = \lim_{\nu\to\infty} \sum_{j=1}^{\nu-1} E[F(t_{j+1})G(t_j) - F(t_j)G(t_{j-1})]$$

$$= \lim_{\nu\to\infty} \sum_{j=1}^{\nu-1} E\left[(F(t_{j+1})-F(t_j))\,\frac{G(t_j)+G(t_{j-1})}{2} + \frac{F(t_{j+1})+F(t_j)}{2}\,(G(t_j)-G(t_{j-1}))\right]$$

$$= \lim_{\nu\to\infty} \sum_{j=1}^{\nu-1} E[(DF(t_j))G(t_j) + F(t_j)D_*G(t_j)]\frac{b-a}{\nu}$$

$$= \int_a^b E[(DF(t))G(t) + F(t)D_*G(t)]\,dt\,.$$

This proves the integration by parts formula (5.19).

But there is another integration by parts formula:

$$(5.20)\quad \int_a^b E(DF(t))G(t)\,dt = -\int_a^b EF(t)DG(t)\,dt - \int_a^b E\sigma^{ij}\nabla_i f(\xi(t))\,\nabla_j g(\xi(t))\,dt$$

$$+ EF(b)G(b) - EF(a)G(a)\,.$$

This is an immediate consequence of the algebraic identity

$$d(F(t)G(t)) = (dF(t))G(t) + F(t)dG(t) + dF(t)dG(t)$$

and the fact that

$$dF(t)dG(t) \equiv \nabla_i f(\xi(t))d\xi^i \nabla_j g(\xi(t))d\xi^j \equiv \sigma^{ij}\nabla_i f(\xi(t))\nabla_j g(\xi(t))\,dt\,.$$

By (5.19) and (5.20),

$$\frac{d}{dt} Ef(\xi(t))g(\xi(t)) = E(Df(\xi(t))g(\xi(t)) + Ef(\xi(t))D_* g(\xi(t))$$

$$= E(Df(\xi(t))g(\xi(t)) + Ef(\xi(t))Dg(\xi(t)) + E\sigma^{ij}\nabla_i f(\xi(t))\nabla_j g(\xi(t))\,,$$

so that

$$(5.21)\quad Ef(\xi(t))D_* g(\xi(t)) = Ef(\xi(t))Dg(\xi(t)) + E\sigma^{ij}\nabla_i f(\xi(t))\nabla_j g(\xi(t))\,.$$

Now $Dg(\xi(t)) = \left(\frac{1}{2}\Delta + b^i\nabla_i\right)g(\xi(t))$ and $D_* g(\xi(t)) = \left(-\frac{1}{2}\Delta_* + b_*^i\nabla_{*i}\right)g(\xi(t))$, where Δ_* is the Laplacian for σ_{*ij}, ∇_* is the Riemannian connection for σ_{*ij}, and b_*^i in normal coordinates for σ_{*ij} is the *backward drift*. Therefore (5.21) is

$$\int f(x)\left[\left(-\frac{1}{2}\Delta_* + b_*^i\nabla_{*i}\right)g(x)\right]\rho(x,t)\,d_M x$$

$$= \int f(x)\left[\left(\frac{1}{2}\Delta + b^i\nabla_i\right)g(x)\right]\rho(x,t)\,d_M x + \sigma^{ij}\nabla_i f(x)\,\nabla_j g(x)\,\rho(x,t)\,d_M x\,.$$

But the last integral, on integration by parts, is

$$-\int f(x)\,\Delta g(x)\,\rho(x,t)\,d_M x - \int f(x)\,\nabla_j g(x)\,\sigma^{ij}\,\frac{\nabla_i\,\rho(x,t)}{\rho(x,t)}\,\rho(x,t)\,d_M x\,.$$

Therefore the *backward diffusion operator* is

$$-\frac{1}{2}\,\Delta_* + b_*^i \nabla_{*i} = -\frac{1}{2}\,\Delta + (b^i - \nabla^i \log \rho)\nabla_i\,.$$

This shows that $\sigma_*^{ij} = \sigma^{ij}$ (so we can drop the $*$ on Δ_* and ∇_*) and $b_*^i = b^i - \nabla^i \log \rho$. This argument, which is more elegant than my exposition of it, is due to Eric Carlen (unpublished), to whom I express my thanks.

We define the *osmotic velocity* u^i by $u^i = (b^i - b_*^i)/2$, so that we have the *osmotic equation*

$$u^i = \frac{1}{2}\,\frac{\nabla^i \rho}{\rho} = \nabla^i \log \sqrt{\rho}\,. \tag{5.22}$$

From the forward Fokker-Planck equation for $\check{\xi}$, which has probability density $\check{\rho}$, we see that ρ also satisfies the *backward Fokker-Planck equation*

$$\frac{\partial \rho}{\partial t} = -\frac{1}{2}\,\Delta\rho - \nabla_i(b_*^i \rho)\,. \tag{5.23}$$

We define the *current velocity* v^i by $v^i = (b^i + b_*^i)/2$. Averaging (5.17) and (5.23), we obtain the *current equation* (also called the equation of continuity)

$$\frac{\partial \rho}{\partial t} = -\nabla_i(v^i \rho)\,. \tag{5.24}$$

The transformation law of $d_*\xi^i$ is

$$d_*\xi^{i'} \equiv \frac{\partial q^{i'}}{\partial q^i}\,d\xi^i - \frac{1}{2}\,\frac{\partial^2 q^{i'}}{\partial q^i \partial q^j}\,\sigma^{ij}\,dt\,. \tag{5.25}$$

We define the *symmetric vector differential*

$$\circ d\xi^i = \frac{d\xi^i + d_*\xi^i}{2} = \frac{\xi^i(t+dt) - \xi^i(t-dt)}{2}. \tag{5.26}$$

Comparing (5.6) and (5.25), we see that $\circ d\xi^i$ is a vector. We have $E_t \circ d\xi^i \equiv v^i dt$.

It may seem odd that $(\xi^i(t+dt) - \xi^i(t-dt))/2$ is a vector but $\xi^i(t+dt) - \xi^i(t)$ is not, but it must be remembered that the time t enters into the definition of the tangency relation $\equiv$.

The stochastic calculus developed in this section may be compared with that of K. Itô in [36]. He considers continuous quasimartingales in R^n, which roughly speaking are processes X with a decomposition $X_t = M_t + A_t$ where M is a continuous martingale and A has continuous paths of locally bounded variation. He introduces a multiplication of stochastic differentials by

$$dX_t \cdot dY_t = d(XY)_t - \int_t^{t+dt} XdY - \int_t^{t+dt} YdX,$$

where the integrals are Itô stochastic integrals (whereas here $d\xi(t)d\eta(t)$ is simply the ordinary product). Theorem 1 of Itô's paper asserts that if X, Y, and Z are continuous quasimartingales, then $dX \cdot dY \cdot dZ = 0$ and $dX \cdot dY$ is locally of bounded variation (pathwise, as dt varies). Despite the differences in the multiplicative structure, (5.3) and (5.2) correspond to these results both in form and application. See also [4] [32] [45].

The term "normal coordinates" was introduced by G. D. Birkhoff in [8, p. 123]. He selects a special class of local coordinates that is defined in terms of the exponential map. This seems to be the standard approach; clearly it requires the Riemannian connection to be already given. The approach used here inverts the order of presentation. For an invariant construction of the Riemannian connection, see [48, §7].

§6. Markovian Diffusion

An M-valued stochastic process is a *Markov process* in case for each t the past and the future are conditionally independent, given the present; that is, if $p \in \widetilde{\mathcal{P}}_t$ and $f \in \widetilde{\mathcal{F}}_t$, then $E_t pf = E_t p E_t f$ whenever the integrals exist.

The notion of a Markov process is invariant under time reversal.

We denote $E\{\cdot|\mathcal{P}_t\}$ by $\overline{E}_t$. Since $\mathcal{P}$ is increasing, we have $\overline{E}_s = \overline{E}_s \overline{E}_t$ for $s \le t$. Let p and f be L^2 functions in $\widetilde{\mathcal{P}}_t$ and $\widetilde{\mathcal{F}}_t$ respectively, and let ξ be a Markov process. Then $f - E_t f \in \widetilde{\mathcal{F}}_t$, so $E(p(f-E_t f)) = EE_t(p(f-E_t f)) = EE_t p E_t(f-E_t f) = 0$. That is, $Epf = E(pE_t f)$, which by definition of conditional expectation means that $\overline{E}_t f = E_t f$. Consequently, $E_s f = E_s E_t f$ for $s \le t$ and $f \in \widetilde{\mathcal{F}}_t$. This is called the *Markov property*. Conversely, a stochastic process with the Markov property is a Markov process.

A *smooth Markovian diffusion* is a smooth diffusion that is a Markov process.

Let ξ be a smooth Markovian diffusion. Let f be in $C_0^\infty(M)$ and let $s < t$. Then $E_s f(\xi(t)) \in \widetilde{\mathcal{N}}_s$, so there is a function $F(x,s)$ (the dependence on t is suppressed in the notation) such that $E_s f(\xi(t)) = F(\xi(s),s)$. By the Markov property and (5.16),

$$F(\xi(s),s) = E_s f(\xi(t)) = E_s E_{s+ds} f(\xi(t)) = E_s F(\xi(s+ds),s+ds)$$
$$= F(\xi(s),s) + \left(\frac{\partial}{\partial s} + b^i \nabla_i + \frac{1}{2}\Delta\right) F(\xi(s),s)\,ds + o(ds)\,,$$

so that F satisfies the *forward diffusion equation* (usually called the backward diffusion equation!)

$$\left(\frac{\partial}{\partial s} + b^i \nabla_i + \frac{1}{2}\Delta\right) F = 0\,. \tag{6.1}$$

For each t, F depends linearly on f, so there exists a unique p, called the *forward transition probability*, such that

$$F(x,s) = \int p(x,s;dy,t)\,f(y)\,.$$

Since f is arbitrary, p satisfies the forward diffusion equation in x and s. By the same argument used in §5 for ρ, for each x and s with $s < t$ there is a C^∞ strictly positive function $p(x,s;y,t)$ such that $p(x,s;dy,t) = p(x,s;y,t)\, d_M y$, and p satisfies the forward Fokker-Planck equation in y and t.

Similarly, we have the *backward transition probability* $p_*(dx,s;y,t)$ with density $p_*(x,s;y,t)$, defined for $s < t$, which satisfies the *backward diffusion equation*, $\left(\frac{\partial}{\partial s} + b_*^i \nabla_i - \frac{1}{2}\Delta\right) G = 0$, in y and t and the backward Fokker-Planck equation in x and s. Since

$$Ef(\xi(s))\, g(\xi(t)) = \iint f(x)\, \rho(x,s)\, p(x,s;y,t)\, g(y)\, d_M x\, d_M y$$

$$= \iint f(x)\, p_*(x,s;y,t)\, \rho(y,t)\, g(y)\, d_M x\, d_M y \,,$$

we have

$$\rho(x,s)\, p(x,s;y,t) = p_*(x,s;y,t)\, \rho(y,t)\,. \tag{6.2}$$

For $t_1 < \cdots < t_n$ we define

(6.3) $\rho(dx_1,t_1;\cdots;dx_n,t_n) =$

$$p_*(dx_1,t_1;x_2,t_2)\cdots p_*(dx_{i-1},t_{i-1};x_i,t_i)\,\rho(dx_i,t_i)\,p(x_i,t_i;dx_{i+1},t_{i+1})\cdots$$

$$\cdots p(x_{n-1},t_{n-1};dx_n,t_n)\,.$$

By (6.2), this is independent of the choice of i. Let $f_1,\cdots,f_n$ be bounded measurable functions on M. By the Markov property,

$$Ef_1(\xi(t_1))\cdots f_n(\xi(t_n)) = \int\cdots\int f_1(x_1)\cdots f_n(x_n)\,\rho(dx_1,t_1;\cdots;dx_n,t_n)\,. \tag{6.4}$$

Some of the factors f_i in (6.4) may be 1 without affecting the left-hand side. Thereby we derive the *Chapman-Kolmogorov equations*

$$\int p(x,r;dy,s)\,p(y,s;z,t) = p(x,r;z,t) \tag{6.5}$$

and similarly for p_*.

The $\rho(dx_1,t_1;\cdots;dx_n,t_n)$ determine the regular probability measure Pr on path space. Since (6.3) is independent of the choice of i, the measure $\rho(dx,r)$ for a fixed r together with $p_*(dx,s;y,t)$ for $s \le t \le r$ and $p(x,s;dy,t)$ for $r \le s \le t$ determine Pr. We may *condition* a smooth Markovian diffusion by replacing $\rho(dx,r)$ for a fixed r by another probability measure $\hat{\rho}(dx,r)$ on M, keeping the same p_* for $s \le t \le r$ and p for $r \le s \le t$, and therefore the same $b_*^i(x,t)$ for $t \le r$ and $b^i(x,t)$ for $t \ge r$. The conditioned probability density $\hat{\rho}$ satisfies the backward Fokker-Planck equation for $s < r$ with $\hat{\rho}(dx,r)$ as final condition and the forward Fokker-Planck equation for $t > r$ with $\hat{\rho}(dy,r)$ as initial condition. The conditioned $\hat{u}^i$ is given by $\hat{u}^i = \frac{1}{2}\nabla^i \log \hat{\rho}$; it is singular at r unless $\hat{\rho}(dz,r)$ has a smooth density $\hat{\rho}(z,r)$. The conditioned b^i for $s < r$ is determined by $\hat{u}^i = (\hat{b}^i - b_*^i)/2$ and the conditioned $\hat{b}_*^i$ for $t > r$ is determined by $\hat{u}^i = (b^i - \hat{b}_*^i)/2$. If $\hat{\rho}(z,r)$ is smooth, then ξ is a smooth Markovian diffusion; otherwise $\hat{\xi}$ is merely a smooth Markovian diffusion on each of $\{s \in I : s \le r\}$ and $\{t \in I : t \ge r\}$.

In applications, ρ is the probability density produced by the physical interaction and $\hat{\rho}(dz,r)$ is the information obtained by looking at the configuration at time r (for example, if I see that $\xi(r) = z_0$, then I take $\hat{\rho}(dz,r)$ to be δ_{z_0}). Since ρ is determined by the physical interaction, it makes sense to ask what dynamical equation it satisfies — but there is no reason to expect $\hat{\rho}$ to satisfy the same dynamical equation.

§7. Continuity of Paths

Here is an example of a Markov process: M is (–1,1) and ξ is the Wiener process except that as soon as the particle reaches ± 1 it is immediately sent to 0, where it starts over. This pathological process has discontinuous paths. It is not a smooth diffusion because our conditions fail to hold for the time reversed process. The Wiener process with reflection at ± 1 is a smooth diffusion, and it has continuous paths in $\dot{M}$. These examples will be useful to bear in mind when reading the following proofs. The proof is easier if M is compact.

THEOREM 7.1. *Let ξ be a smooth Markovian diffusion on* M, *let* Pr *be the corresponding regular probability measure on path space* Ω, *and let* Z *be the set of all continuous paths* ω *from* I *to* $\dot{M}$. *Then* $\Pr(Z) = 1$.

Proof. The manifold M has a metric r (for example, the metric constructed from σ_{ij}). It will be convenient to define $r(\infty,x) = \infty$ for all x in M. For any subset B of M, let

$$N_\varepsilon B = \{y \in M : r(y,x) \leq \varepsilon \text{ for some } x \text{ in } B\}$$

and let $N_\varepsilon^c B$ be its complement. Let K be a compact subset of M, let $\varepsilon > 0$ be small enough so that $N_{2\varepsilon}K$ is compact, and let J be a compact subinterval of I. For x_0 in K, let f be a smooth positive function that is 0 on $N_{\varepsilon/2}\{x_0\}$ and 1 on $N_\varepsilon^c\{x_0\}$. Then

$$\frac{d}{dt} E_t f(\xi(t)) = \left(\frac{1}{2}\Delta + b^i \nabla_i\right) f(\xi(t)). \tag{7.1}$$

Let $g(x,t) = \left(\frac{1}{2}\Delta + b^i(x,t)\nabla_i\right) f(x)$. Then

$$\frac{d^2}{dt^2} E_t f(\xi(t)) = \left(\frac{1}{2}\Delta + b^i \nabla_i + \frac{\partial}{\partial t}\right) g(\xi(t),t). \tag{7.2}$$

Both (7.1) and (7.2) are 0 for $\xi(t)$ in $N_{\varepsilon/2}\{x_0\}$, so that $E_t f(\xi(t+dt)) =$

$o(dt^2)$ for $\xi(t)$ in $N_{\varepsilon/2}\{x_0\}$. (In fact, it is $o(dt^n)$ for all n.) But

$$p(x,t;\, N^c_\varepsilon\{x\},t+dt) \le p(x,t;\, N^c_{\varepsilon/2}\{x_0\},t+dt) \le E_t f(\xi(t+dt)) = o(dt^2)$$

for $x = \xi(t)$ in $N_{\varepsilon/2}\{x_0\}$. Let

$$\theta(K,J,\varepsilon,\delta) = \sup\{p(x,s;\, N^c_\varepsilon\{x\},t) : x \in K, s \in J, t \in J, t-s \le \delta\}\,.$$

Then, by compactness, $\theta(K,J,\varepsilon,\delta) = o(\delta^2)$. Let $\theta_1(K,J,\varepsilon,\delta)$ be the supremum of the sums $\Sigma\, \theta(K,J,\varepsilon,\delta_i)$ where $\Sigma\, \delta_i \le \delta$; then $\theta_1(K,J,\varepsilon,\delta) = o(\delta)$. Let $t_1 < \cdots < t_n$ be in J with $t_n - t_1 \le \delta$, and let

$$A = \{\omega : r(\omega(t_1),\omega(t_j)) > \varepsilon \text{ for some } j = 1,\cdots,n\}\,.$$

Let $\Pr_{x,t}$ be the probability measure on path space conditioned by $\xi(t) = x$. I claim that for x in K,

$$\Pr_{x,t_1}(A) \le 2\theta\left(N_{2\varepsilon}K,J,\tfrac{1}{2}\varepsilon,\delta\right) + \theta_1(N_\varepsilon K,J,\varepsilon,\delta)\,. \tag{7.3}$$

To see this, let

$$B = \left\{\omega : r(\omega(t_1),\omega(t_n)) > \tfrac{1}{2}\varepsilon\right\},$$

$$C_j = \left\{\omega : r(\omega(t_j),\omega(t_n)) > \tfrac{1}{2}\varepsilon\right\},$$

$$D_j = \{\omega : 2\varepsilon \ge r(\omega(t_1),\omega(t_j)) > \varepsilon \text{ and } r(\omega(t_1),\omega(t_k)) \le \varepsilon \text{ for all } k = 1,\cdots,j-1\}\,,$$

$$D = \{\omega : r(\omega(t_{j-1}),\omega(t_j)) > \varepsilon \text{ for some } j = 2,\cdots,n\}\,.$$

By the triangle inequality,

$$A \subseteq B \cup \bigcup_{j=2}^{n} (C_j \cap D_j) \cup E\,.$$

(If ω is at some time ε-far from its initial value $[\omega \in A]$, then either the final value is $\frac{1}{2}$ ε-far from the initial value $[\omega \in B]$ or there is a first j such that $\omega(t_j)$ is ε-far from the initial value and $\frac{1}{2}$ ε-far from the final value. But either $\omega(t_j)$ is ε-far from $\omega(t_{j-1})$ $[\omega \in E]$ or $\omega(t_j)$ is not 2ε-far from the initial value $[\omega \in C_j \cap D_j]$.)

Now $\Pr_{x,t_1}(B) \leq \theta\left(K,J,\frac{1}{2}\varepsilon,\delta\right)$ and $\Pr_{x,t_1}(E) \leq \theta_1(N_\varepsilon K,J,\varepsilon,\delta)$ for x in K. Observe that $C_j \in \mathcal{F}_{t_j}$ and $D_j \in \mathcal{P}_{t_j}$. Let $\Pr_{x,t;t_j}$ be the conditional probability for $\mathcal{N}_{t_j}$ with respect to $\Pr_{x,t}$; then

$$\Pr_{x,t;t_j}(C_j \cap D_j) = \Pr_{x,t;t_j}(C_j)\Pr_{x,t;t_j}(D_j) \leq \theta\left(N_{2\varepsilon}K,J,\frac{1}{2}\varepsilon,\delta\right)\Pr_{x,t;t_j}(D_j)$$

for x in K. (If this is unclear, write out the integrals using (6.4).) But the D_j are disjoint, so $\sum_j \Pr_{x,t_1}(C_j \cap D_j) \leq \theta\left(N_{2\varepsilon}K,J,\frac{1}{2}\varepsilon,\delta\right)$ for x in K. This proves (7.3).

Abbreviate the right-hand side of (7.3) by $\theta_2(K,J,\varepsilon,\delta)$. Let K^0 be the interior of K and let

$$F = \{\omega : r(\omega(t_j),\omega(t_k)) > 2\varepsilon \text{ and } \omega(t_j) \in K^0 \text{ for some } j \text{ and } k \text{ with } 1 \leq j < k \leq n\}$$

I claim that

$$\Pr(F) \leq \theta_2(K,J,\varepsilon,\delta)\ . \tag{7.4}$$

To see this, let

$$G_i = \{\omega : \omega(t_i) \in K^0 \text{ and } \omega(t_\ell) \notin K^0 \text{ for all } \ell = 1,\cdots,i-1\}\ ,$$

$$H_i = \{\omega : r(\omega(t_i),\omega(t_j)) > \varepsilon \text{ for some } j = i,\cdots,n\}\ .$$

Then $F \subseteq \bigcup_i (G_i \cap H_i)$, the G_i are disjoint, $G_i \in \mathcal{P}_{t_i}$, $H_i \in \mathcal{F}_{t_i}$, and H_i is of the form A with $1,\cdots,n$ replaced by $i,\cdots,n$, so that (7.3) applies to it. Therefore

$$\Pr(F) \leq \sum_i \Pr(G_i) \sup_{x \in K} \Pr_{x,t_i}(H_i) \leq \theta_2(K,J,\varepsilon,\delta)\ ,$$

which proves (7.4). Now apply this result to the time reversed process. Let F_* be defined as F but with $1 \le j < k \le n$ replaced by $1 \le k < j \le n$. Then there is a $\theta_{2*}(K,J,\varepsilon,\delta)$, which is $o(\delta)$, such that $\Pr(F_*) \le \theta_{2*}(K,J,\varepsilon,\delta)$. Let $L = F \cup F_*$, and let $\theta_3(K,J,\varepsilon,\delta) = \theta_2(K,J,\varepsilon,\delta) + \theta_{2*}(K,J,\varepsilon,\delta)$. Then $\Pr(L) \le \theta_3(K,J,\varepsilon,\delta)$. Let $[a,b] \subseteq J$ with $b-a \le \delta$, and let

$$Q = \{\omega: r(\omega(s), \omega(t)) > 2\varepsilon \text{ and } \omega(s) \in K^0 \text{ for some } s \text{ and } t \text{ in } [a,b]\}.$$

I claim that

$$\Pr(Q) \le \theta_3(K,J,\varepsilon,\delta). \tag{7.5}$$

To see this, let $S = \{t_1, \cdots, t_n\}$ with $a \le t_1 < \cdots < t_n \le b$, and denote our previous L by L_S. Then L_S is an open set, the union of any two such is a set of the same form, and $Q = \cup L_S$. By Theorem 3.1 we have (7.5).

Let k be the least integer greater than $|J|/\delta$, where $|J|$ is the length of J, and partition J into k subintervals of length $\le \delta$. If s and t are in J with $|s-t| \le \delta$, then they lie in the same or adjacent subintervals. Let $\theta_4(K,J,\varepsilon,\delta) = k\theta_3(K,J,\varepsilon,\delta)$, so that $\theta_4(K,J,\varepsilon,\delta) = o(1)$, and let

$$R = \{\omega: r(\omega(s),\omega(t)) > 4\varepsilon \text{ and } \omega(s) \in K^0 \text{ for some } s \text{ and } t \text{ in } J \text{ with } |s-t| \le \delta\}. \tag{7.6}$$

By (7.5), $\Pr(R) \le \theta_4(K,J,\varepsilon,\delta)$. (If s and t lie in adjacent intervals, let r be the common endpoint. Then $r(\omega(s),\omega(r)) > 2\varepsilon$ or $r(\omega(r),\omega(t)) > 2\varepsilon$.)

Now let K_n be a sequence of compact sets whose interiors cover M. Let J_ℓ be a sequence of compact subintervals whose union is I. The set R of (7.6) depends on K, J, ε, and δ. Then

$$Z = \bigcap_n \bigcap_\ell \bigcap_{\varepsilon>0} \bigcup_{\delta>0} R^c(K_n, J_\ell, \varepsilon, \delta)$$

and we may restrict ε and δ to be rational. Since each $R^c(K_n, J_\ell, \varepsilon, \delta)$

is a closed set, this shows that Z is a Borel set, so that $\Pr(Z)$ is defined. To show that $\Pr(Z) = 1$, we need only show for each fixed n, ℓ, and ε that $\Pr(R(K_n, J_{\ell,\varepsilon}, \delta)) \to 0$ as $\delta \to 0$, but this is precisely what we have shown. ■

§8. Stochastic Integrals

Let ξ be a smooth Markovian diffusion, and let A_i be a possibly time-dependent covector field. We want to give meaning to the expression

$$\int A_i(\xi(t),t)\,\dot{\xi}^i\,dt \,. \tag{8.1}$$

Let A_i be smooth and have support in a coordinate neighborhood U, with local coordinates $q^1, \cdots, q^n$. We define w^i by

$$\xi^i(t) - \xi^i(s) = \int_s^t \beta^i(\xi(r),r)\,dr + w^i(t) - w^i(s)\,; \tag{8.2}$$

this determines w^i up to an additive constant that may be chosen arbitrarily (better yet, regard $w^i(t) - w^i(s)$ as a "difference process" indexed by $I \times I$; see [47, p. 83]). Notice that the integral in (8.2) is an ordinary Riemann integral (by the continuity of paths). So also is $\int_s^t A_i(\xi(r),r)\,\beta^i(\xi(r),r)\,dr$.

By (8.2),

$$d\xi^i(t) \equiv \beta^i(\xi(t),t)\,dt + dw^i(t)\,.$$

Recall that $E_t d\xi^i \equiv \beta^i dt$, so that $E_t dw^i \equiv 0$ and $dw^i dw^j \equiv d\xi^i d\xi^j \equiv \sigma^{ij} dt$. (If $M = \mathbf{R}^n$ with $\sigma^{ij} = \delta^{ij}$, then w^i is the Wiener process.)

Since $\frac{d}{dt} E_s \xi^i(t) = E_s \beta^i(\xi(t),t)$, we have

$$E_s \xi^i(t) = \xi^i(s) + E_s \int_s^t \beta^i(\xi(t),r)\,dr\,. \tag{8.3}$$

By (8.2) and (8.3), $E_s[w^i(t)-w^i(s)] = 0$, so that w^i is a martingale. Thus (8.2) is the continuous time analogue of the decomposition (4.4) of a discrete time process into a trend process and a martingale. Stochastic integrals are the analogue of sums (4.8) giving investment strategies, but I will discuss them only for the special case that the integrand is in $\widetilde{\mathcal{P}}_t$.

Now divide the interval $[s,t]$ into ν equal pieces: $r_\alpha = s+\alpha(t-s)/\nu$ for $\alpha = 0, \cdots, \nu$. Let $dr_\alpha = (t-s)/\nu$ and $dw^i(r_\alpha) = w^i(r_\alpha + dr_\alpha) - w^i(r_\alpha)$. I claim that

$$\text{(8.4)} \qquad \int_s^t A_i(\xi(r),r)\,dw^i(r) = \lim_{\nu\to\infty} \sum_{\alpha=0}^{\nu-1} A_i(\xi(r_\alpha),r_\alpha)\,dw^i(r_\alpha)$$

exists in L^2. To see this, consider more generally

$$\text{(8.5)} \qquad J = \sum_{\alpha-0}^{\nu-1} \eta_i(r_\alpha)\,dw^i(r_\alpha)$$

where each $\eta_i(r_\alpha)$ is in $\widetilde{\mathcal{P}}_{r_\alpha} \cap L^\infty$. We want to compute EJ^2. Notice that for $\alpha < \beta$ we have

$$E\eta_i(r_\alpha)\,dw^i(r_\alpha)\,\eta_j(r_\beta)\,dw^j(r_\beta) = E\eta_i(r_\alpha)\,dw^i(r_\alpha)\,\eta_j(r_\beta)\,E_{r_\beta}dw^j(r_\beta) = 0\,,$$

so only the terms $\alpha = \beta$ contribute. Thus

$$EJ^2 = E\sum_{\alpha=0}^{\nu=1} \eta_i(r_\alpha)\,\eta_j(r_\alpha)\,dw^i(r_\alpha)\,dw^j(r_\alpha)$$

$$= E\sum_{\alpha=0}^{\nu-1} \eta_i(r_\alpha)\,\eta_j(r_\alpha)\,\sigma^{ij}(\xi(r_\alpha))\,dr_\alpha + o(1)\,.$$

Now if we compare the sums in (8.4) for two different values ν_1 and ν_2, their difference may be written in the form (8.5) for $\nu = \nu_1\nu_2$ (a common refinement of the partitions) with $\eta_i(r_\alpha)$ that is a difference of values of $A_i(\xi(r),r)$ for two values of r differing by at most $(b-a)/\min(\nu_1,\nu_2)$. By the continuity of paths and the Lebesgue dominated convergence theorem, $EJ^2 \to 0$ as $\nu_1, \nu_2 \to \infty$, which proves the claim.

We define

$$(8.6)\quad \int_s^t A_i(\xi(r),r)\,d\xi^i(r) = \int_s^t A_i(\xi(r),r)\,\beta^i(\xi(r),r)\,dr + \int_s^t A_i(\xi(r),r)\,dw^i(r)\,.$$

This depends on the choice of local coordinates, because $d\xi^i$ is not a vector and so $A_i d\xi^i$ is not a scalar. The two integrals

$$(8.7)\quad \int_s^t A_i(\xi(r),r)\,\tilde{d}\xi^i(r) = \int_s^t A_i(\xi(r),r)\,d\xi^i(r) + \int_s^t A_i(\xi(r),r)\,\frac{1}{2}\Gamma^i(\xi(r))\,dr\,,$$

$$(8.8)\quad \int_s^t A_i(\xi(r),r)\,\tilde{d}_*\xi^i(r) = \int_s^t A_i(\xi(r),r)\,d_*\xi^i(r) - \int_s^t A_i(\xi(r),r)\frac{1}{2}\Gamma^i(\xi(r))\,dr\,,$$

where $\tilde{d}_*\xi^i(r) = d_*\xi^i(r) - \frac{1}{2}\Gamma^i(\xi(r)dr$ is the *backward vector differential*, do not depend on the choice of local coordinates, but the use of either one involves the choice of a direction of time.

Recall that $\circ d\xi^i = \frac{1}{2}\,d\xi^i + \frac{1}{2}\,d_*\xi^i$ is a vector, and define

$$(8.9)\quad \int_s^t A_i(\xi(r),r)\circ d\xi^i(r) = \frac{1}{2}\int_s^t A_i(\xi(r),r)\,d\xi^i(r) + \frac{1}{2}\int_s^t A_i(\xi(r),r)\,d_*\xi^i(r)\,.$$

This is independent of the choice of local coordinates and of the direction of time; we use it to give meaning to the formal expression (8.1).

The integral (8.7) is an Itô integral [33] in the sense of Meyer [45], and (8.9) is a Fisk [25]-Stratonovich [59] integral in the sense of Ikeda and Manabe [31]. The use of the symmetric vector differential in path integrals of covector fields is originally due to Feynman [23].

§9. Stochastic Action

Let ξ be a smooth Markovian diffusion and consider the formal expression

$$E\int \frac{1}{2}\,\dot{\xi}^i\dot{\xi}_i\,dt\,. \tag{9.1}$$

Can we attach any meaning to this? Since $d\xi^i d\xi_i \equiv \sigma_i^i dt = ndt$, where n is the dimension of M, this looks hopelessly singular: $\int \frac{n}{2}$, with no differential! But let us examine it more closely.

Let $\tilde{d}\xi$ be the vector in $T_{\xi(t)}M$ such that a free particle starting at $\xi(t)$ at time t, and moving with constant velocity $\tilde{d}\xi/dt$, arrives at $\xi(t+dt)$ at time $t+dt$. (In §5, $\tilde{d}\xi$ was defined only up to $o(dt)$, as an element of $\mathcal{B}_t/\mathcal{O}_t$.) Then $\tilde{d}\xi^i\tilde{d}\xi_i$ is the square of the distance between $\xi(t+dt)$ and $\xi(t)$. We want to compute $E_t\tilde{d}\xi^i\tilde{d}\xi_i$ up to $o(dt^2)$.

Let η be the trajectory of the freely moving particle, so that

$$\eta^i(t) = \xi^i(t)\,,$$

$$\eta^i(t+dt) = \xi^i(t+dt)\,,$$

$$\left(\frac{d}{ds}\,\eta^i\right)(t) = \frac{\tilde{d}\xi^i}{dt}\,,$$

and

$$\frac{d^2\eta^i}{ds^2} + \Gamma^i_{jk}\frac{d\eta^j}{ds}\frac{d\eta^k}{ds} = 0$$

since the acceleration of η is 0. Choose normal coordinates q^ℓ at $\xi(t)$, and abbreviate $\partial/\partial q^\ell$ by ∂_ℓ. When the argument of a function is not displayed, it is understood to be $\xi(t)$. Then

$$\left(\frac{d^2\eta^i}{ds^2}\right)(t) = 0\,,$$

$$\left(\frac{d^3\eta^i}{ds^3}\right)(t) = -\partial_\ell\Gamma^i_{jk}\,\frac{\tilde{d}\xi^j}{dt}\,\frac{\tilde{d}\xi^k}{dt}\,\frac{\tilde{d}\xi^\ell}{dt}\,,$$

so that

$$d\xi^i = \eta^i(t+dt) - \eta^i(t) = \tilde{d}\xi^i - \frac{1}{6}\,\partial_\ell \Gamma^i_{jk} \tilde{d}\xi^j \tilde{d}\xi^k \tilde{d}\xi^\ell + O(dt^2)\,,$$

$$\tilde{d}\xi^i = d\xi^i + \frac{1}{6}\,\partial_\ell \Gamma^i_{jk} d\xi^j d\xi^k d\xi^\ell + O(dt^2)\,.$$

Therefore

$$\tilde{d}\xi^i \tilde{d}\xi_i = d\xi^i d\xi_i + \frac{1}{3}\,\partial_\ell \Gamma^i_{jk} d\xi^j d\xi^k d\xi^\ell d\xi_i + o(dt^2)\,,$$

$$(9.2)\quad E_t \tilde{d}\xi^i \tilde{d}\xi_i = E_t d\xi^i d\xi_i + \frac{1}{3}\,\delta^{jk}\,\partial_i \Gamma^i_{jk} dt^2 + \frac{2}{3}\,\delta^{\ell k}\,\partial_\ell \Gamma^i_{ik} dt^2 + o(dt^2)\,.$$

The numerators 1 and 2 come from the fact that i may be paired with ℓ and i may be paired with j or k. Notice that the term with $i = j = k = \ell$ occurs in both terms, but $E_t d\xi^i d\xi^i d\xi^i d\xi_i = 3dt^2 + o(dt^2)$.

By (8.2),

$$d\xi^i = \int_t^{t+dt} \beta^i(\xi(r),r)\,dr + dw^i\,.$$

The integral is $b^i dt + o(dt)$, since $\beta^i = b^i$ at $\xi(t)$ in normal coordinates, but we need a better approximation to compute $d\xi^i d\xi_i$ up to $o(dt^2)$, because dw^i is of order $dt^{1/2}$. Apply (8.2) to $\xi(r)$ in the integrand; then

$$d\xi^i = \int_t^{t+dt} \beta^i\left(\xi(t) + \int_t^r \beta(\xi(s),s)\,ds + w(r) - w(t), r\right) dr + dw^i$$

$$= b^i dt + \partial_k \beta^i W^k + dw^i + O(dt^2)$$

where

$$W^k = \int_t^{t+dt} [w^k(r) - w^k(t)]\,dr\,.$$

By (5.14), this gives

$$d\xi^i = b^i dt + \nabla_k b^i w^k - \frac{1}{2}\delta^{j\ell}\partial_k \Gamma^i_{jk} w^k + dw^i + O(dt^2),$$

so that

$$(9.3)\quad d\xi^i d\xi_i = b^i b_i dt^2 + 2b^i dw_i dt + 2\nabla_k b^i w^k dw_i - \delta^{j\ell}\partial_k \Gamma^i_{j\ell} w^k dw_i + dw^i dw_i + o(dt^2)$$

The term $2b^i dw_i dt$ is singular, but

$$E_t 2b^i dw_i dt = 0 .$$

The process w has orthogonal increments, so if $t \le s \le r$ then

$$E_t[w^k(s) - w^k(t)][w_i(r) - w_i(t)] = E_t \int_t^s dw^k(s_1) \int_t^r \sigma_{ia}(\xi(s_2))\, dw^a(s_2)$$

$$= E_t \int_t^s dw^k(s_1) \int_t^r \left[\delta_{ia} + \frac{1}{2}\delta^{j\ell}\partial_j\partial_\ell \sigma_{ia}(s_2 - t)\right] dw^a(s_2) + o(s-t)^2$$

$$= \delta^k_i(s-t) + \frac{1}{4}\delta^{ka}\delta^{j\ell}\partial_j\partial_\ell\sigma_{ia}(s-t)^2 + o(s-t)^2 .$$

Therefore

$$E_t 2\nabla_k b^i w^k dw_i = \nabla_i b^i dt^2 + o(dt^2),$$

$$E_t(-\delta^{j\ell}\partial_k \Gamma^i_{j\ell} w^k dw_i)) = -\frac{1}{2}\delta^{j\ell}\partial_i \Gamma^i_{j\ell} dt^2 + o(dt^2),$$

$$E_t dw^i dw_i = n dt + \frac{1}{4}\delta^{ia}\delta^{j\ell}\partial_j\partial_\ell\sigma_{ia} dt^2 + o(dt^2) .$$

Since $\nabla_\ell \sigma_{ia} = 0$ —that is, $\partial_\ell \sigma_{ia} + 2\Gamma^b_{\ell i}\sigma_{ba} = 0$ —we have $\partial_j\partial_\ell\sigma_{ia} = -2\partial_j(\Gamma^b_{\ell i}\sigma_{ba})$, so that in normal coordinates

$$\frac{1}{4}\delta^{ia}\delta^{j\ell}\partial_j\partial_\ell\sigma_{ia} = -\frac{1}{2}\delta^i_b\delta^{j\ell}\partial_j\Gamma^b_{\ell i} = -\frac{1}{2}\delta^{\ell k}\partial_\ell \Gamma^i_{ik} .$$

By (9.2) and (9.3),

$$E_t \tilde{d}\xi^i \tilde{d}\xi_i = \left(b^i b_i + \nabla_i b^i - \frac{1}{6}\delta^{jk}\partial_i \Gamma^i_{jk} + \frac{1}{6}\delta^{\ell k}\partial_\ell \Gamma^i_{ik}\right) dt^2 + n dt + o(dt^2) .$$

But in normal coordinates, $-\frac{1}{6}\delta^{jk}\partial_i\Gamma^i_{j\ell} + \frac{1}{6}\delta^{\ell k}\partial_\ell \Gamma^i_{ik} = \frac{1}{6}\bar{R}$, where $\bar{R}$ is the scalar curvature. Therefore in general coordinates

(9.4) $$E_t \frac{1}{2}\frac{\tilde{d}\xi^i}{dt}\frac{\tilde{d}\xi_i}{dt} = \frac{1}{2} b^i b_i + \frac{1}{2}\nabla_i b^i + \frac{1}{12}\bar{R} + \frac{n}{2dt} + o(1) .$$

This result is due to Guerra (unpublished). After seeing the manuscript draft of this section, which was wrong from beginning to end, he kindly informed me of the calculations that he had previously made, leading to (9.4).

The singular term $\frac{n}{2dt}$ is a constant that is the same for all paths, and it drops out in the variation of the expected kinetic action. We will use this fact in Chapter II.

§10. Stochastic Parallel Translation

In deterministic kinematics, the idea of parallel translation along a path ξ is that a neighboring configuration η moves so as to have the same velocity at all times (see the concluding remarks in §2). Let us try to generalize this idea to stochastic kinematics.

Let ξ be a smooth diffusion. Does it make sense to say that at all times $d\eta^i(t)$ is the same as $d\xi^i(t)$? No, because they are not vectors—but it does make sense to say that at all times $\tilde{d}\eta^i(t)$ is the same as $\tilde{d}\xi^i(t)$.

Let the neighboring configuration $\eta(t)$ be defined by $\eta^i = \xi^i + Y^i$ where Y^i is a tangent vector at $\xi(t)$. Then

(10.1) $$\Gamma^i_{k\ell}(\eta(t)) = \Gamma^i_{k\ell} + \frac{\partial \Gamma^i_{k\ell}}{\partial q^m} Y^m + o(Y) .$$

Christoffel symbols and their derivatives are understood to be evaluated at $\xi(t)$ unless an argument is displayed.

The condition for the neighboring vectors $\tilde{d}\xi^i$ and $\tilde{d}\eta^i$ to be the same is

$$\tilde{d}\eta^i = \tilde{d}\xi^i - \Gamma^i_{k\ell} Y^k \tilde{d}\xi^\ell . \tag{10.2}$$

We have

$$d\xi^i \equiv \tilde{d}\xi^i - \frac{1}{2}\Gamma^i_{k\ell}\tilde{d}\xi^k\tilde{d}\xi^\ell , \tag{10.3}$$

$$d\eta^i = \tilde{d}\eta^i - \frac{1}{2}\Gamma^i_{k\ell}(\eta(t))\tilde{d}\eta^k\tilde{d}\eta^\ell , \tag{10.4}$$

$$dY^i \equiv d\eta^i - d\xi^i . \tag{10.5}$$

Putting these five equations together, we find

$$dY^i \equiv -\Gamma^i_{k\ell}Y^k d\xi^\ell - \frac{1}{2}\left(\frac{\partial\Gamma^i_{k\ell}}{\partial q^m} - \Gamma^i_{kj}\Gamma^j_{\ell m} + \Gamma^i_{j\ell}\Gamma^j_{km}\right) Y^m \sigma^{k\ell} dt . \tag{10.6}$$

This is a linear equation in Y, and its solution for an initial value $Y^i(s)$ is of the form

$$Y^i(t) = \tau_\xi(s,t)^i_j Y^j(s) ,$$

where $\tau_\xi(s,t) : T_{\xi(s)} \to T_{\xi(t)}$. We call τ_ξ *stochastic parallel translation.* It is due to Dohrn and Guerra, and is in general different from a parallel translation introduced by Itô. The induced map on tensor spaces will also be denoted by $\tau_\xi(s,t)$.

Let a be a smooth tensor field of compact support. Now we can define its *stochastic forward derivative* by

$$Da(\xi(t)) = \lim_{dt\to 0+} E_t \frac{\tau_\xi(t+dt,t)\, a(\xi(t+dt)) - a(\xi(t))}{dt} . \tag{10.7}$$

To compute this, consider the case that a is a vector field Y^i and choose normal coordinates at $\xi(t)$. Then

$$\tau_\xi(t,t+dt)\,Y^i(\xi(t)) \equiv Y^i(\xi(t)) - \frac{1}{2}\,\frac{\partial\Gamma^i_{k\ell}}{\partial q^m}\,Y^m\sigma^{k\ell}dt\ ,$$

so that

$$\tau_\xi(t+dt,t)\,Y^i(\xi(t+dt)) \equiv Y^i(\xi(t+dt)) + \frac{1}{2}\,\frac{\partial\Gamma^i_{k\ell}}{\partial q^m}\,Y^m\sigma^{k\ell}dt\ .$$

Then

$$\tau_\xi(t+dt,t)\,Y^i(\xi(t+dt)) - Y^i(\xi(t)) \equiv [Y^i(\xi(t+dt)) - Y^i(\xi(t))] +$$

$$\frac{1}{2}\,\frac{\partial\Gamma^i_{k\ell}}{\partial q^m}\,Y^m\sigma^{k\ell}dt \equiv \frac{\partial Y^i}{\partial q^j}\,d\xi^j + \frac{1}{2}\,\frac{\partial^2 Y^i}{\partial q^j\partial q^p}\,\sigma^{jp}dt + \frac{1}{2}\,\frac{\partial\Gamma^i_{k\ell}}{\partial q^m}\,Y^m\sigma^{k\ell}dt\ ,$$

and thus

$$DY^i = b^j\frac{\partial Y^i}{\partial q^j} + \frac{1}{2}\,\sigma^{jp}\,\frac{\partial^2 Y^i}{\partial q^j\partial q^p} + \frac{1}{2}\,\sigma^{k\ell}\,\frac{\partial\Gamma^i_{k\ell}}{\partial q^m}\,Y^m \quad (NC)\ .$$

But in normal coordinates $b^j\dfrac{\partial Y^i}{\partial q^j} = b^j\nabla_j Y^i$ and

$$\frac{1}{2}\,\sigma^{jp}\nabla_j\nabla_p Y^i = \frac{1}{2}\,\sigma^{jp}\nabla_j\left(\frac{\partial Y^i}{\partial q^p} + \Gamma^i_{pm}Y^m\right)$$

$$= \frac{1}{2}\,\sigma^{jp}\,\frac{\partial^2 Y^i}{\partial q^j\partial q^p} + \frac{1}{2}\,\sigma^{jp}\,\frac{\partial\Gamma^i_{pm}}{\partial q^j}\,Y^m\ .$$

Consequently,

$$DY^i = b^j\nabla_j Y^i + \frac{1}{2}\,\sigma^{jp}\nabla_j\nabla_p Y^i + \frac{1}{2}\,\sigma^{k\ell}\left(\frac{\partial\Gamma^i_{k\ell}}{\partial q^m} - \frac{\partial\Gamma^i_{km}}{\partial q^\ell}\right)Y^m \quad (NC)\ .$$

But in normal coordinates, the expression in parentheses is the Riemann curvature tensor $R^i{}_{km\ell}$, by (2.11), and $\sigma^{k\ell}R^i{}_{km\ell}$ is minus the Ricci tensor R^i_m. Therefore

$$(10.8) \qquad DY^i = \frac{1}{2}(\nabla^j\nabla_j Y^i - R^i_j Y^j) + b^j\nabla_j Y^i .$$

This is a tensor equation, and is valid in general coordinates. For a general tensor field a we have, using the notation of (1.5),

$$Da = \frac{1}{2}(\nabla^j\nabla_j a - R^{\cdot}_{\cdot} a) + b^j\nabla_j a .$$

We denote the *Dohrn-Guerra Laplacian* $\nabla^j\nabla_j - R^{\cdot}_{\cdot}$ simply by Δ. Then for possibly time dependent tensor fields, and with the definition of the backward stochastic derivative D_* for tensor fields analogous to that of D, our equations have the same appearance as for scalars:

$$(10.9) \qquad \begin{cases} Da = \left(\frac{1}{2}\Delta + b^j\nabla_j + \frac{\partial}{\partial t}\right) a , \\ D_* a = \left(-\frac{1}{2}\Delta + b^j_*\nabla_j + \frac{\partial}{\partial t}\right) a . \end{cases}$$

A very useful fact is that Δ commutes with grad and div. Let f be a smooth scalar; then I claim that

$$(10.10) \qquad \nabla_i\Delta f = \Delta\nabla_i f .$$

This follows from Ricci's identity (2.12). We have

$$\nabla_i\nabla_j\nabla_k f = \nabla_j\nabla_i\nabla_k f - R^a{}_{kij}\nabla_a f = \nabla_j\nabla_k\nabla_i f - R^a{}_{kij}\nabla_a f ,$$

and contracting with σ^{jk} we find

$$\nabla_i\Delta f = \nabla^a\nabla_a\nabla_i f + R^a_i\nabla_a f = \Delta\nabla_i f .$$

We may write (10.10) as grad $\Delta = \Delta$ grad. Since Δ is formally self-adjoint, it also commutes with the formal adjoint (–div) of grad, so that

$$(10.11) \qquad \nabla_i\Delta X^i = \Delta\nabla_i X^i .$$

On scalars and 1-forms, the Dohrn-Guerra Laplacian agrees with (minus!) the Laplace-de Rham operator $-(d\delta+\delta d)$, but this is not true in general for forms of higher degree.

We define $D\xi^i = b^i(\xi(t),t)$ and $D_*\xi^i = b^i_*(\xi(t),t)$, and we define the *stochastic acceleration* a^i by

$$a^i(\xi(t),t) = \frac{1}{2}(D_*D+DD_*)\xi^i . \tag{10.12}$$

Other reasonable definitions, such as $\frac{1}{2}(DD+D_*D_*)\xi^i$ or any average of this with (10.12), are possible, but we will see later that (10.12) enters the theory naturally — it no longer has to be postulated as in [47]. We have

$$a^i = \frac{1}{2}D_*b^i+\frac{1}{2}Db^i_* = \frac{1}{2}\left(-\frac{1}{2}\Delta+b^j_*\nabla_j+\frac{\partial}{\partial t}\right)b^i+\frac{1}{2}\left(\frac{1}{2}\Delta+b^j\nabla_j+\frac{\partial}{\partial t}\right)b^i_* \tag{10.13}$$

$$= \left(\frac{\partial v^i}{\partial t}+v^j\nabla_j v^i\right)-\left(\frac{1}{2}\Delta u^i+u^j\nabla_j u^i\right).$$

A notion of stochastic parallel translation was introduced by Itô in [34]; see also [19], [35], and [45]. Itô's construction has the property that if the time interval is subdivided into a sequence of partitions with mesh tending to 0, then the ordinary parallel translations along the piecewise-geodesic approximations to the path converge to it with probability one. The Itô translation is isometric. The stochastic parallel translation described here is due to Dohrn and Guerra [14][15]. A neighboring point traveling the same way as the diffusing particle feels the Ricci curvature: on a positively curved manifold it tends to get closer. Dohrn and Guerra call this geodesic deviation.

§11. Existence of Diffusions

The simplest example of a smooth Markovian diffusion is the *Wiener process* w. Here $M = \mathbf{R}^n$, $\sigma^{ij} = \delta^{ij}$, $I = [0,\infty)$, $\rho(dx,0) = \delta$, $b^i = 0$. The probability measure Pr is given by (6.4) with

$$(11.1)\qquad p(x,s;y,t) = (2\pi(t-s))^{-\frac{n}{2}} e^{-\frac{(x-y)^2}{2(t-s)}},\quad 0 \le s < t < \infty .$$

This is the stochastic analogue of rest, and books have been written about it. The w^i are independent Gaussian processes, and $w^i(t) - w^i(s)$ for $s \le t$ is Gaussian with mean 0 and variance $t-s$. To verify that w is a smooth diffusion, we need to show that $dw^i dw^j \equiv \delta^{ij} dt$. Let z be the difference of the two sides. Then $E_t z = 0$ and $E_t z^2 = dt^2$ for $i \neq j$ and $E_t z^2 = 2dt^2$ for $i = j$ (since $E_t(dw^i)^4 = 3dt^2$). Thus $z \in \mathcal{O}_t$. I will not start talking about the Wiener process because if I did, I might not stop.

The Wiener process may be used to construct a general smooth Markovian diffusion locally. Consider the process conditioned to be at x at time s. Choose local coordinates at x in M. We want to construct, locally, a Markov process for given β^i and σ^{ij}. Suppose we can solve

$$(11.2)\qquad d\xi^i(t) \equiv \beta^i(\xi(t),t)\,dt + A^i_a(\xi(t))\,dw^a$$

where w is the Wiener process. Then $E_t d\xi^i \equiv \beta^i dt$ and $d\xi^i d\xi^j \equiv A^i_a A^j_b \delta^{ab} dt$, so we want $A^i_a A^j_b \delta^{ab} = \sigma^{ij}$. To achieve this, let A^i_a be the square root of positive type of the matrix $\sigma^{ij}\delta_{ja}$ (this is a coordinate-dependent construction). With this choice of A^i_a, let us proceed to solve (11.2). This can be done in strict analogy with the deterministic case, Theorem 1.1. Let $\xi^i_0(t) = x^i$ and inductively

$$\xi^i_{n+1}(t) = x^i + \int_s^t \beta^i(\xi_n(r),r)\,dr + \int_a^t A^i_a(\xi_n(r))\,dw^a(r) .$$

The second integral is a stochastic integral. The convergence proof is quite analogous to the proof of Theorem 1.1. Then the limiting process has the required properties so long as it stays in the coordinate neighborhood.

The motions in different coordinate neighborhoods may be glued together to construct the process until (or if) it runs off the manifold to ∞; see [26]. What one gets is an analogue of Theorem 1.2. As the counterexample at the beginning of §7 shows, we need to know more than β^i and σ^{ij} to tell the particle what to do if it runs off the manifold.

Another local construction is the analogue of the deterministic polygonal approximation method (see §1). Choose a random vector $\tilde{d}w^i(s)$ in T_x of mean 0 and covariance $\sigma^{ij}(x)ds$; for example, it may be Gaussian or uniformly distributed over the sphere of radius $\sqrt{ds}$. Let $\tilde{d}\xi^i(s) = b^i(x,s)ds + \tilde{d}w^i(s)$. Then let the particle travel with constant velocity $\tilde{d}\xi^i(s)/ds$ for time ds, arriving at a point $\xi^i(s+ds)$. In contrast to the deterministic case, it is essential to use the Riemannian connection to define what is meant by traveling with constant velocity (acceleration vector 0). Then choose a $\tilde{d}w^i(s+ds)$ in $T_{\xi(s+ds)}$ independently, and repeat. This construction is harder to control than the method of successive approximations, but it is coordinate-independent and very appealing geometrically. The use of nonstandard analysis produces substantial simplifications in this kind of construction. The polygonal approximation describes the motion as long as the particle stays in M.

These two methods are direct probabilistic constructions. Another approach is to use methods of partial differential equations to solve for the forward and backward transition probabilities and then construct the measure on path space by (6.4).

Let us begin the discussion with the assumption that the Riemannian manifold M is compact, so as not to have to worry about boundary conditions. Suppose that at a fixed time, say 0, a probability measure $\rho(dx,0)$ on M is given, and a smooth vector field $b^i(x,t)$ is given for all $t \geq 0$. Then for x in M and $0 \leq s$, the forward Fokker-Planck equation (in y and t, for $s < t$) with the Dirac measure at x as initial condition at time s, has a unique smooth strictly positive solution

$p(x,s;y,t)$, which is probability density in y. The uniqueness ensures that it satisfies the Chapman-Kolmogorov equation (6.5). Let $I = [0,\infty)$ and let Ω be path space. For f in $C_{\mathrm{fin}}(\Omega)$, of the form (3.2), let

$$(11.3)\quad L(f) = \int\cdots\int F(x_1,\cdots,x_n)\,\rho(dx_0,0)\,p(x_0,0;dx_1,t_1)\,p(x_1,t_1;dx_2,t_2)\cdots p(x_{n-1},t_{n-1};dx_n,t_n)\,.$$

Then (6.5) ensures that (11.3) is well-defined (the F corresponding to f is not unique, since there may be some t_i such that f does not depend on $\omega(t_i)$). Let Pr be the corresponding regular probability measure on Ω, as in §3, and let $\xi(t)$ be the random variable $\omega \mapsto \omega(t)$. Then ξ is a smooth diffusion.

This is not hard to see. The σ^{ij} and b^i are given. For f a smooth scalar, verify that

$$(11.4)\qquad Df(\xi(t)) = \left(\frac{1}{2}\Delta + b^i\nabla_i\right) f(\xi(t))\,.$$

As a consequence, if g is also a smooth scalar,

$$(11.5)\qquad D(fg) = (Df)\,g + f(Dg) + \sigma^{ij}\nabla_i f\,\nabla_j g\,.$$

Use (11.4) and (11.5) for f and g local coordinate functions to verify (5.1) and (5.2). Then consider the time reversed process. Notice that the probability density $\rho(x,t) = \int \rho(dx_0,0)\,p(x_0,0;x,t)$ is strictly positive and smooth, and define b_*^i by the osmotic equation (5.22). Use (6.2) as a definition of p_* and observe that it can equally well be used to construct the measure on path space (on $(0,t)$). Use this to verify that $D_*f(\xi(t)) = \left(-\frac{1}{2}\Delta + b_*^i\nabla_i\right)f(\xi(t))$, and argue as before.

For our purposes, rather than prescribe ρ at one time and b^i at all future times, it is more natural to prescribe ρ at all times. Let M still be a compact Riemannian manifold, and let ρ be a strictly positive smooth function on $M \times \mathbf{R}$ that for each t is a probability density. Is there a smooth diffusion on M with ρ as probability density? Some people seem to think that in a diffusion process ρ must spread out as time increases, but this is not so. It can spread out, bunch up again, spread out, bunch up at five separated peaks, produce rings of alternating density – in short, there are no restrictions. The reason we don't observe this kind of behavior when ink diffuses in water is a dynamical one – this diffusion has dissipative dynamics. But the motion of billiard balls is also less interesting if someone has poured molasses all over the table.

The osmotic velocity $u^i = \frac{1}{2}\nabla^i \log \rho$ is smooth since $\rho > 0$. We need to find a current velocity satisfying the current equation (5.24). This is easily done: let

$$v^i = -\frac{1}{\rho}\nabla^i \Delta^{-1}\frac{\partial\rho}{\partial t}.$$

Notice that $\Delta^{-1}\frac{\partial\rho}{\partial t}$ exists since

$$\int_M \frac{\partial\rho}{\partial t}\, d_M x = \frac{d}{dt}\int_M \rho d_M x = \frac{d}{dt} 1 = 0.$$

There are other solutions, obtained by adding $\rho^{-1}z^i$ where z^i is any smooth divergence-free vector field. Now for any smooth v^i satisfying the current equation, let $b^i = v^i + u^i$ and $b_*^i = v^i - u^i$. Then ρ satisfies the forward Fokker-Planck equation (5.17) since

$$\frac{1}{2}\Delta\rho = \nabla_i\left(\frac{1}{2}\nabla^i\rho\right) = \nabla_i(u^i\rho),$$

and similarly it satisfies the backward Fokker-Planck equation (5.23). Let p and p_* be the fundamental solutions of these two equations; then they are related by (6.2). Define $\rho(dx_1,t_1;\cdots;dx_n,t_n)$ by (6.3), for f in $C_{fin}(\Omega)$ of the form (3.2) let

$$L(f) = \int\cdots\int F(x_1,\cdots,x_n)\,\rho(dx_1,t_1;\cdots,dx_n,t_n),$$

let Pr be the corresponding probability measure on Ω, and let $\xi(t)$ be $\omega \mapsto \omega(t)$. Then ξ is a smooth diffusion with probability density ρ.

Now let us consider a general Riemannian manifold M, beginning with the case that $v^i = 0$, so that by the current equation ρ is constant in time. I call this *osmotic diffusion*; see [1] and [2]. Let us make the assumption of finite osmotic energy,

$$\int \frac{1}{2} u^i u_i \rho d_M < \infty . \tag{11.6}$$

(It is not really an energy; the dimensionality is off by a factor with the dimensions of action, but that will be introduced later.)

Let $\mathcal{H}$ be the real Hilbert space $L^2(M, \rho d_M x)$, let $\mathcal{E}$ be the subspace of all smooth f that are constant outside a compact set, let

$$\|f\|_1^2 = \frac{1}{2}\int \nabla^i f \nabla_i f \rho d_M x + \int f^2 \rho d_M x , \tag{11.7}$$

and let $\mathcal{H}^1$ be the completion of $\mathcal{E}$ in this norm. I claim that the quadratic form (11.7) is closeable (so that we may identify $\mathcal{H}^1$ with a dense subspace of $\mathcal{H}$). Let f_n be a sequence in $\mathcal{E}$ such that $f_n \to \phi$ in $\mathcal{H}^1$ and $f_n \to 0$ in $\mathcal{H}$; we need to show that $\phi = 0$. For all g in $\mathcal{E}$, recalling that $\frac{1}{2}\nabla^i \rho = u^i \rho$ we find

$$\frac{1}{2}\int \nabla^i f_n \nabla_i g \rho d_M x = -\int f_n \frac{1}{2} \Delta g \rho d_M x - \int f_n \nabla_i g u^i \rho d_M x \to 0$$

by (11.6). Hence $(\phi, g) = \lim_n (f_n, g) = 0$, so that $\phi = 0$.

Now we define $\frac{1}{2}\Delta_\rho$ to be the negative self-adjoint operator on $\mathcal{H}$ corresponding to this quadratic form. The operator

$$e^{t\frac{1}{2}\Delta_\rho}$$

is an integral operator whose kernel (with respect to the measure $\rho d_M x$) we denote by $p^t(x,y)$. This may be used to construct Pr on path space. Notice that 1 is in the domain of Δ_ρ and $\Delta_\rho 1 = 0$.

This variational approach has the advantage not only of specifying good boundary conditions (Neumann condition—reflection at the boundary) but of working with minimal smoothness assumptions. Heretofore we have considered only smooth diffusions, but we will be interested in more general cases. In particular, the construction above allows ρ to have *nodes* where $\rho = 0$ (and u^i becomes infinite).

The general diffusion is a continual mixture of osmotic diffusion and *current flow* (the flow generated by the current velocity). I was convinced that one could prove existence and uniqueness theorems for solutions of the diffusion equation, and construct Pr on path space, under the assumption that for all $s \le t$,

$$\int_s^t \int_M \frac{1}{2}(u^i u_i + v^i v_i)\rho d_M x\, dr < \infty, \tag{11.8}$$

without smoothness or strict positivity assumptions. Eric Carlen, as part of his Princeton thesis, has discovered how to do this, using variational methods from the theory of partial differential equations; see [67].

Chapter II
DYNAMICS OF CONSERVATIVE DIFFUSION

The first two sections of this chapter are a review of deterministic dynamics. It is pointed out that the requirement that the force be a dynamical veriable restricts the class of allowable Lagrangians.

In §14 conservative diffusion processes are characterized directly in terms of the ordinary Lagrangian, and the Schrödinger equation is derived. This gives a direct dynamical prescription, with no arbitrariness, for the diffusion processes of Markovian stochastic mechanics. This section is entitled "Stochastic quantization", but the physics remains classical.

The final section of this chapter concerns the nodes (zeros) of the wave function, showing that a unique Markovian diffusion process is associated with a general solution of the Schrödinger equation (with some smoothness assumptions but no restrictions on the nodes), and discussing the probabilistic weighting to be assigned when the nodes divide space-time into noncommunicating regions.

§12. Newtonian Dynamics

Let F be a force acting on a mechanical system whose configuration is x, and let $X^i dt$ be an infinitesimal displacement of the configuration (so that $X \in T_x M$). Then the work done by the force is $dW = F_i X^i dt$. The much-maligned tensor notation, with its profusion of upper and lower indices, shows immediately that a force is a cotangent vector; for dW to be independent of the choice of local coordinates, the components F_i must transform according to

$$F_{i'} = \frac{\partial q^i}{\partial q^{i'}} F_i \, .$$

a force, we are assuming that the local coordinates q^i sions of length. Sometimes it is convenient to choose local coordinates with other dimensions, and the different components need not all have the same dimensions. But dW should always have the dimensions of energy. When the q^i do not all have the dimensions of length, but $F_i X^i dt$ has the dimensions of energy, we should properly speak of F as being a "generalized force.") A *force field* is a dynamical variable whose value at each x is a cotangent vector at x, so $F: TM \times \mathbf{R} \to T^*M$ with $\pi F(x,v,t) = x$.

In Newtonian mechanics the fundamental dynamical law is $F = ma$. As we have seen, we need an affine connection in order to define the acceleration a. Which affine connection should we choose? And after we have chosen it, how do we set a covector equal to m times a vector? What is the mass m?

These problems are solved by introducing the notion of the *kinetic energy* of a mechanical system. The kinetic energy T, a smooth scalar on velocity phase space TM, is given in local coordinates by

$$T = \frac{1}{2} m_{ij} p^i p^j .$$

For this to be independent of the choice of local coordinates, m_{ij} must be a covariant 2-tensor, and since $p^i p^j = p^j p^i$, without loss of generality we take m_{ij} to be symmetric. The kinetic energy is always positive, and is strictly positive unless $p^i = 0$. In short, m_{ij} is a Riemannian metric on configuration space M. If the q^i have the dimensions of length, then m_{ij} has the dimensions of mass; we call it the *mass tensor*.

We choose the affine connection ∇ to be the Riemannian connection, and use it to define the acceleration: $a^i = \ddot{\xi}^i + \Gamma^i_{jk} \dot{\xi}^j \dot{\xi}^k$. Then Newton's equation is $F_i = m_{ij} a^j$. For a given force field F we obtain equations of motion for a local flow on velocity phase space; in local coordinates,

$$\dot{q}^i = p^i ,$$
$$\dot{p}^i = m^{ij} F_j - \Gamma^i_{jk} p^j p^k .$$

§13. Lagrangian Dynamics

The Newtonian formulation of dynamics is too general — it ignores the overriding importance of energy conservation in an isolated mechanical system. Let us recall the more familiar Lagrangian formulation of dynamics. I will argue that the Lagrangian formulation is also too general, and that the proper formulation of dynamics is the common part of the two formulations.

Let $L : TM \times R \to R$ be smooth, with the dimensions of energy; it is called a *Lagrangian*. (For an isolated system, L will be independent of the time t.) Given a path ξ in the configuration space M, with velocity vector $\dot{\xi}$, we define

$$I = \int_{t_0}^{t_1} L(\xi, \dot{\xi}, t)\, dt .$$

Hamilton's principle is that the path ξ is a critical point of I, under variations with the same initial and final points. How does one vary a path in a manifold? Let X be a time dependent vector field on M, and recall that $\xi^i(t) + X^i(\xi(t), t)$ gives a neighboring point $\eta^i(t)$ to $\xi^i(t)$, which is well-defined (independently of the choice of local coordinates) up to $o(X)$. We abbreviate $X^i(\xi(t), t)$ by $\delta\xi^i$ and $\frac{d}{dt} X^i(\xi(t), t)$ by $\delta\dot{\xi}^i$. For a functional A of paths, let $\delta A(\xi)$ be $A(\xi + \delta\xi) - A(\xi)$, to first order in X. Recall the derivation of the *Euler-Lagrange equations*:

$$\delta I = \int_{t_0}^{t_1} [L(\xi + \delta\xi, \dot{\xi} + \delta\dot{\xi}, t) - L(\xi, \dot{\xi}, t)]\, dt$$

$$= \int_{t_0}^{t_1} \left[\frac{\partial L}{\partial q^i}\delta\xi^i + \frac{\partial L}{\partial p^i}\delta\dot{\xi}^i\right] dt = \int_{t_0}^{t_1} \left[\frac{\partial L}{\partial q^i} - \frac{d}{dt}\frac{\partial L}{\partial p^i}\,\delta\xi^i\right] dt ,$$

under the assumption that $X(\cdot, t_o) = X(\cdot, t_1) = 0$, so that

$$\frac{\partial L}{\partial q^i} - \frac{d}{dt}\frac{\partial L}{\partial p^i} = 0 . \tag{13.1}$$

Now let M be a Riemannian manifold with mass tensor m_{ij}, kinetic energy $T = \frac{1}{2} m_{ij} p^i p^j$, and define the potential energy V by

$$L = T - V .$$

Then $\frac{\partial L}{\partial q^i} = \frac{\partial T}{\partial q^i} - \frac{\partial V}{\partial q^i}$, so that in normal coordinates at a point,

$$\frac{\partial L}{\partial q^i} = -\frac{\partial V}{\partial q^i} \quad \text{(NC)}.$$

By the Euler-Lagrange equation (13.1),

$$0 = -\frac{\partial V}{\partial q^i} - \frac{d}{dt}\left(\frac{\partial T}{\partial p^i} - \frac{\partial V}{\partial p^i}\right) = -\frac{\partial V}{\partial q^i} - \frac{d}{dt}\left(m_{ij}p^j - \frac{\partial V}{\partial p^i}\right) \quad \text{(NC)} .$$

But by Newton's equation, $\frac{d}{dt}(m_{ij}p^j) = F_i$ (NC), so that we have

$$F_i = -\frac{\partial V}{\partial q^i} + \frac{d}{dt}\frac{\partial V}{\partial p^i} \quad \text{(NC)} .$$

But we require F_i *to be a dynamical variable.* If the state in TM of a mechanical system is known, and the time is known, then the force F must be known. By this requirement, $\frac{\partial V}{\partial p^i}$ must be independent of p; that is, V(x,p,t) must be of the form

$$V(x,p,t) = \phi(x,t) - A_j(x,t)p^j . \tag{13.2}$$

This is a tensor equation, so it holds in general coordinates. We call ϕ the *scalar potential* and A_i the *covector potential.* Then

$$F_i = \left(-\frac{\partial \phi}{\partial q^i} - \frac{\partial A_i}{\partial t}\right) + \left(\frac{\partial A_j}{\partial q^i} - \frac{\partial A_i}{\partial q^j}\right)p^j .$$

We set

$$E_i = -\frac{\partial\phi}{\partial q^i} - \frac{\partial A_i}{\partial t} \tag{13.3}$$

and call it the *electric-type force*, and we set

$$H_{ij} = \frac{\partial A_j}{\partial q^i} - \frac{\partial A_i}{\partial q^j}\,; \tag{13.4}$$

it is the exterior derivative of the convector potential A, and we call $H_{ij}p^j$ the *magnetic-type force*. It is always orthogonal to the velocity: $H_{ij}p^ip^j = 0$ since $H_{ij} = -H_{ji}$.

Let $\chi : M \times \mathbf{R} \to \mathbf{R}$ be smooth, and make the replacements

$$\begin{cases} \phi \to \phi + \dfrac{\partial\chi}{\partial t} \\ A_i \to A_i - \dfrac{\partial\chi}{\partial q^i} \end{cases} \tag{13.5}$$

in the Lagrangian. Then the equations of motion are the same after this *gauge transformation.*

By a *basic dynamical system* I mean a Riemannian manifold M and a Lagrangian $L = T - V$, where $T = \frac{1}{2} m_{ij}p^ip^j$ and V is of the form (13.2).

A familiar example of a basic dynamical system is given by a particle of mass m in a scalar potential V. Here $M = \mathbf{R}^3$ and $L = \frac{1}{2} mv^2 - V(x)$. This is a nonrelativistic system. The relativistic kinetic energy of a particle is not expressed by a Riemannian metric, and we cannot write a relativistic analogue of this system as a basic dynamical system. But there is no relativisitic dynamics of interacting particles; relativisitic dynamics requires fields, and relativistic fields can be represented as basic dynamical systems in the limit as the number n of degrees of freedom goes to infinity.

Now let us change our point of view towards the principle of least action. Rather than follow the motion of a single configuration, let us study the flow on M determined by a time dependent velocity field v. Let $\xi(s;x,t)$ be the configuration at time s if it is x at time t. Let L be a basic Lagrangian, $L = \frac{1}{2} m_{ij}p^i p^i - \phi + A_i v^i$, and define *Hamilton's principal function* $S(x,t)$ by

$$S(x,t) = -\int_t^{t_1} L(\xi(s;x,t), v(\xi(s;x,t),s),s)\,ds .$$

(This function of x and t depends also on t_1, and it is a functional of v.) Let D be the substantial derivative (derivative along paths), so that $D = \frac{\partial}{\partial t} + v^i \nabla_i$. Then $DS = L$; that is, $DS(x,t) = L(x,v(x,t),t)$.

Now we require that v be a critical point of S, for unconstrained variations. Let v' be another time dependent vector field, let $\delta v = v' - v$, and denote the quantities with v' replacing v by $'$. Then

$$\begin{aligned} D(S'-S) &= D'S' - DS + (D-D')S' = L' - L - \delta v^i \nabla_i S' \\ &= L' - L - \delta v^i \nabla_i S + o(\delta v) . \end{aligned}$$

Now $L' - L = (v_i + A_i)\delta v^i + o(\delta v)$, and $S' - S = -\int_t^{t_1} D(S'-S)\,ds$ since S' and S vanish at t_1, so that

$$S' - S = -\int_t^{t_1} (v^i + A_i - \nabla_i S)\,\delta v^i ds + o(\delta v) .$$

Since this is $o(\delta v)$ for all choices of δv^i, we have the *Hamilton-Jacobi condition*

(13.6) $$v_i + A_i = \nabla_i S ,$$

which together with $DS = L$ – i.e. $\frac{\partial S}{\partial t} + v^i \nabla_i S = \frac{1}{2} v^i v_i - \phi + A_i v^i$ – gives the *Hamilton-Jacobi equation*

$$\frac{\partial S}{\partial t} + \frac{1}{2}(\nabla^i S - A^i)(\nabla_i S - A_i) + \phi = 0 . \tag{13.7}$$

Using (13.6), we may rewrite (13.7) as

$$\frac{\partial S}{\partial t} + \frac{1}{2} v^j v_j + \phi = 0 .$$

Now apply ∇_i and use (13.6) again. Then

$$\frac{\partial v_i}{\partial t} + \frac{\partial A_i}{\partial t} + \frac{1}{2} \nabla_i (v^j v_j) + \nabla_i = 0 . \tag{13.8}$$

For any smooth vector field z^j we have

$$\frac{1}{2} \nabla_i (z^j z_j) = z^j \nabla_j z_i + z^j (\nabla_i z_j - \nabla_j z_i) , \tag{13.9}$$

as may be verified in normal coordinates. Since ∇ is torsion-free (the Christoffel symbols are symmetric in their lower indices), $\nabla_i z_j - \nabla_j z_i$ is the exterior derivative of z_j. By (13.6), v_j has the same exterior derivative as $-A_j$, so (13.8) is

$$\frac{\partial v_i}{\partial t} + v^j \nabla_j v_i = - \nabla_i \phi - \frac{\partial A_i}{\partial t} + H_{ij} v^j . \tag{13.10}$$

Since the acceleration vector a^i is given by Dv^i, (13.10) is Newton's equation $m_{ij} a^j = F_i$, so the second form of the principle of least action leads to the same equation of motion as the first.

Notice that a different choice of final time t_1 in the definition of S replaces S by $S + \chi$ with $D\chi = 0$, which produces a gauge transformation (13.5) in (13.6) and (13.7) and leaves (13.10) unchanged.

The derivation of the Hamilton-Jacobi equation given here is patterned after [28], with a modification in the treatment of the final time.

§14. Stochastic Quantization

Consider a basic dynamical system of finitely many degrees of freedom, with configuration space M and mass tensor m_{ij}. Let us explore the background field hypothesis, that interaction with the background field causes the system to undergo a diffusion process with diffusion tensor $\sigma^{ij} = \hbar m^{ij}$ satisfying a variational principle $\delta E \int L dt = 0$.

This hypothesis is advanced not as a fundamental law of nature, or in an attempt to construct a new kind of mechanics, but as a falsifiable conjecture about the classical interaction of some systems with some background fields.

What kind of system and background field could possibly produce such a diffusion process? An obvious necessary requirement is the dimensional consideration that it be possible to construct a constant with the dimensions of action from the constants of the theory. From this we see that quantum fluctuations are not of gravitational origin: one cannot construct a constant with the dimensions of action from the gravitational constant G and the speed of light c. As is very well known, this can be done from c and the fundamental charge e. I conclude that quantum fluctuations may be of electromagnetic origin (using electromagnetism in a broad sense, to include gauge theories unifying electromagnetic forces with weak and strong forces — a topic that I am not competent to discuss). A major problem is to discover how c and e determine the scale $\hbar$ of the quantum fluctuations.

The kind of interaction with a background that I have in mind is nothing abstruse. Consider, for example, the classical interaction of charged point particles with the electromagnetic field. With suitable ultraviolet and infrared cutoffs, this is a dynamical system of finitely many degrees of freedom, and we have global existence and uniqueness. This is a deterministic, nonlocal system (the motion of a particle in one place immediately affects each of the finitely many field oscillators, and so immediately affects the cutoff field everywhere). Is it an exaggeration to say that nothing whatever is known about the behavior of this system as the cutoffs are removed, that there is not one single theorem that has been proved? The self-interaction of the charged particles is usually ignored or assumed to be a mass renormalization; perhaps, however, as the cutoffs are removed the self-interaction leads to random fluctuations – perhaps the limiting theory is nondeterministic and local. In recent years we have seen various surprising types of chaotic behavior emerge from the dynamics of nonlinear interactions; perhaps quantum fluctuations also are of dynamical origin.

I want to argue that the hypothesis that a system performs a diffusion with diffusion tensor $\hbar m^{ij}$ is just on the borderline of being demonstrably false. Consider a free particle of mass m, and let $\xi(t)$ be its x-coordinate at time t. We may observe $\xi(t)$ to within an error δx, by using light of wave length $\approx \delta x$ or a diaphragm with a slit of width $\approx \delta x$, but as Heisenberg [30] showed, any such measurement gives the particle an uncontrolled additional momentum δp subject to the uncertainty relation $\Delta p \Delta x \geq \hbar/2$, where Δp and Δx are the standard deviations of δp and δx. Let us make the usual assumptions in discussions of this sort, that δp and δx are independent with mean 0 and that the position of the disturbed particle at a later time $t+dt$ is $\xi(t+dt) + \frac{\delta p}{m} dt + o(dt)$. We may observe this later position with arbitrarily great accuracy, since we are not concerned with any disturbance of the motion after time $t+dt$. Then we have observed $\xi(t+dt) - \xi(t)$ to within an error $\frac{\delta p}{m} dt - \delta x$, and so $(\xi(t+dt) - \xi(t))^2$ to within an expected error of $\frac{(\Delta p)^2}{m^2} dt^2 + (\Delta x)^2$. The minimum value of this error, subject to the constraint $\Delta p \Delta x \geq \hbar/2$, is $\frac{\hbar}{m} dt$ –the postulated value of $E_t d\xi(t)^2$. Were one to postulate a bigger value, the only way to avoid a refutation by experiment would be to abandon the "usual assumptions" made above.

Suppose that the system undergoes a diffusion with diffusion tensor $\hbar m^{ij}$ and has probability density ρ at a given time r. We need a dynamical law that, for a given force field, determines the forward drift at all future times and the backward drift at all former times.

First it is necessary to say what this dynamical law will not be. In the review of deterministic dynamics in the two previous sections, we discussed conservative dynamics but said nothing about dissipative dynamics. Suppose we drop a sponge from the leaning tower of Pisa. It will at once acquire a certain velocity, and fall with this constant velocity. If we simultaneously drop a heavier sponge, it will fall with a larger constant velocity and reach the ground first. This is dissipative, or Aristotelian, dynamics: the velocity of a configuration in a force field F is F/β, where β is the coefficient of friction (in tensor notation, $\dot{\xi}^i = \beta^{ij}F_j$). Before Galileo, this was thought to be a fundamental law, rather than an approximate asymptotic law for motion in a resisting fluid, and this misconception made it difficult for many to accept the emerging conservative dynamics. There is an extensive dynamical theory of dissipative diffusion processes: Brownian motion, thermal fluctuations, Einstein-Smoluchowski theory (in which the drift is simply $\beta^{ij}F_j$), Ornstein-Uhlenbeck theory, etc. If you are familiar with dissipative diffusion, forget what you know – it has no more connection with conservative diffusion than has the fall of a sponge with celectial mechanics. In particular, dissipative dynamics – unlike conservative dynamics – chooses a direction of time. If you are strong enough to give a cannonball minus its final velocity, it will rise right back up to where Galileo dropped it; if you give a sponge minus its final velocity, it simply flops to the ground.

We will derive the drift from a variational principle $\delta E\int L dt = 0$. If L is time independent, such a variational principle implies the existence of a conserved energy $T+V$, as is familiar in the deterministic case and as we will verify later in the stochastic case (where the energy is conserved in the sense that its expected value is a constant in time). In terms of the background field hypothesis, we are studying those random motions of the system for which there is on the average no transfer of energy between the system and the background field (but the mutual transfer of energy that averages out to zero is responsible for the quantum fluctuations of the system).

Let me remark that I have no evidence for the background field hypothesis (if I did, I would gladly sacrifice an ox). This hypothesis is in no way used in the mathematical development of stochastic mechanics, but I believe it to be essential for a physical understanding of the theory.

Let ξ be a smooth diffusion on M with diffusion tensor $\sigma^{ij} = \hbar m^{ij}$. We will take the mass tensor m_{ij}, rather than $\sigma_{ij} = \hbar^{-1} m_{ij}$, as the Riemannian metric on M. This has the consequence that some of the formulas of Chapter I acquire factors of $\hbar$. We say that ξ has *finite energy* in case (11.5) holds for all intervals $[s,t]$.

Let us seek the stochastic analogue of the second form of the principle of least action (in which the velocity field is varied). Let

$$L = \frac{1}{2} p^i p_i - \phi + A_i p^i \tag{14.1}$$

where ϕ and A_i are smooth with compact support. For any interval $[t,t_1]$ and natural number ν, let $s_\alpha = t + \alpha(t_1 - t)/\nu$ for $\alpha = 0, \cdots, \nu$, let $ds_\alpha = (t_1 - t)/\nu$, let $d\xi(s_\alpha) = \xi(s_\alpha + ds_\alpha) - \xi(s_\alpha)$, let $s_\alpha^o = (s_{\alpha-1} + s_\alpha)/2$, and let $\tilde{d}\xi(s_\alpha)$ be defined as in §9. We say that a smooth Markovian diffusion ξ is *critical for* L in case it has finite energy and for any interval $[t,t_1]$, whenever δb^i is a smooth time dependent vector field with compact support in $M \times [t,t_1]$, and ξ' is the smooth Markovian diffusion with forward drift $b'^i = b^i + \delta b^i$ and the same diffusion tensor and probability density at time t as ξ,

$$(14.2) \quad \lim_{\nu\to\infty} \Bigg\{ E \sum_{\alpha=1}^{\nu} \left[\frac{1}{2} \frac{\tilde{d}\xi^i(s_\alpha)}{ds_\alpha} \frac{\tilde{d}\xi_i(s_\alpha)}{ds_\alpha} ds_\alpha - \phi(\xi(s_\alpha^o), s_\alpha^o)\, ds_\alpha + A_i(\xi(s_\alpha^o), s_\alpha^o)\, d\xi^i(s_\alpha) \right.$$

$$\left. - E \sum_{\alpha=1}^{\nu} \left[\frac{1}{2} \frac{\tilde{d}\xi'^i(s_\alpha)}{ds_\alpha} \frac{\tilde{d}\xi'_i(s_\alpha)}{ds_\alpha} ds_\alpha - \phi(\xi'(s_\alpha^o), s_\alpha^o)\, ds_\alpha + A_i(\xi'(s_\alpha^o), s_\alpha^o)\, d\xi'^i(s_\alpha) \right] \right\} = o(\delta b$$

Define $\overline{L}_+ : M \times R \to R$ by

$$\overline{L}_+ = \frac{1}{2} b^i b_i + \frac{\hbar}{2} \nabla_i b^i + \frac{\hbar^2}{12} \overline{R} - \phi + A_i b^i + \frac{\hbar}{2} \nabla_i A^i .$$

Then

$$E\overline{L}_+(\xi(t),t) = E\left[\frac{1}{2} \frac{d\xi^i(t)}{dt} \frac{d\xi_i(t)}{dt} - \phi(\xi(t),t) + A_i(\xi(t),t) \circ \frac{d\xi^i(t)}{dt}\right] - \frac{n}{2dt} + o(1) .$$

To see this, use Guerra's formula (9.4) and recall that

$$EA_i \circ d\xi^i(t) = EA_i v^i dt + o(dt) ,$$

but

$$\text{(14.3)} \quad EA_i v^i = \int A_i(x,t)\, v^i(x,t)\, \rho(x,t)\, d_M x = \int A_i b^i \rho - \int A_i u^i \rho$$

$$= \int A_i b^i \rho - \int A_i \frac{\hbar}{2} \nabla^i \rho = \int A_i b^i \rho + \frac{\hbar}{2} \int (\nabla_i A^i) \rho = E\left(A_i b^i + \frac{\hbar}{2} \nabla_i A^i\right) .$$

We define

$$\overline{I} = E \int_t^{t_1} \overline{L}_+(\xi(s),s)\, ds .$$

Then (14.2) is equivalent to

$$\overline{I}' - \overline{I} = o(\delta b) ,$$

where $'$ denotes replacing ξ by ξ'.

The term $\frac{\hbar^2}{12} \overline{R}$ is the Pauli-DeWitt term; see [72] and [73]. There is no consensus as to whether it is spurious. Guerra and Morato [28] do not include it. They observe that

$$E\left(\frac{1}{2} b^i b_i + \frac{\hbar}{2} \nabla_i b^i\right) = E\left(\frac{1}{2} v^i v_i - \frac{1}{2} u^i u_i\right) = E\left(\frac{1}{2} \frac{d\xi^i}{dt} \frac{d_*\xi_i}{dt}\right)$$

and take this as the kinetic action term. In the applications that we will make, it does not matter whether the Pauli-DeWitt term is included or not, because the only curved manifold we will consider is the group SO(3) for which $\bar{R}$ is a constant. Also, in the following discussion we may absorb this term into the scalar potential. My expectation is that the term will prove not to be spurious. If the background field hypothesis is correct, then the motion is governed by a classical Lagrangian for the system plus background field, and $E\left(\frac{1}{2}\frac{\tilde{d}\xi^i}{dt}\frac{\tilde{d}\xi_i}{dt}\right)$ should arise rather than $E\left(\frac{1}{2}\frac{d\xi^i}{dt}\frac{d_*\xi_i}{dt}\right)$.

Following Guerra and Morato [28], let

$$(14.4) \qquad L_+ = \frac{1}{2}b^ib_i + \frac{\hbar}{2}\nabla_i b^i - \phi + A_i b^i + \frac{\hbar}{2}\nabla_i A^i ,$$

$$(14.5) \qquad I = E\int_t^{t_1} L_+(\xi(s),s)\,ds ,$$

and say that ξ is *critical for* L *in the sense of Guerra and Morato* (GM-critical for L) in case

$$(14.6) \qquad I' - I = o(\delta b) .$$

(Thus ξ is GM-critical for L if and only if it is critical for L with ϕ replaced by $\phi - \frac{\hbar^2}{12}\bar{R}$.)

Let $E_{x,t}$ be the conditional expectation with respect to $\mathcal{N}_t$ for the process conditioned by $\xi(t) = x$, and define

$$(14.7) \qquad S(x,t) = -E_{x,t}\int_t^{t_1} L_+(\xi(s),s)\,ds .$$

This is the stochastic analogue of Hamilton's principal function. By (14.7),

(14.8) $$DS = L_+ .$$

Therefore

(14.9) $$D(S'-S) = D'S'-DS + (D-D')S' = L'_+ - L_+ - \delta b^i \nabla_i S'$$
$$= L'_+ - L_+ - \delta b^i \nabla_i S + o(\delta b) .$$

Now

(14.10) $$L'_+ - L_+ = (b_i + A_i)\delta b^i + \frac{\hbar}{2}\nabla_i \delta b^i + o(\delta b) .$$

Since S and S' vanish at t_1, and ρ and ρ' are the same at time t,

(14.11) $$-E\int_t^{t_1} D(S'-S)\,ds = ES'(\xi(t),t) - ES(\xi(t),t) = E'S'(\xi'(t),t) - ES(\xi(t),t)$$
$$= -I' + I .$$

By (14.9), (14.10), and (14.11),

(14.12) $$I' - I = E\int_t^{t_1} \left(b_i + A_i - \nabla_i S + \frac{\hbar}{2}\nabla_i\right)\delta b^i ds + o(\delta b) .$$

But

(14.13) $$E\,\frac{\hbar}{2}\,\nabla_i \delta b^i(\xi(s),s) = \int_M \frac{\hbar}{2}\,(\nabla_i \delta b^i)\rho = -\int_M \delta b^i u_i \rho .$$

Since $b_i - u_i = v_i$,

(14.14) $$I' - I = E\int_t^{t_1} (v_i + A_i - \nabla_i S)\,\delta b^i ds + o(\delta b) .$$

Therefore (14.6) is equivalent to the *stochastic Hamilton-Jacobi condition*

(14.15) $$v_i + A_i = \nabla_i S ,$$

since we may take $\delta b^i = f(x)(v^i + A^i - \nabla^i S)$ where f is any smooth function of compact support in M.

We have proved the following theorem.

THEOREM 14.1. *Let* L *be given by (14.1), where* ϕ *and* A_i *are smooth with compact support. Then a smooth Markovian diffusion* ξ *of finite energy is critical for* L *in the sense of Guerra and Morato if and only if the stochastic Hamilton-Jacobi condition holds.*

Let us continue to assume that ξ is GM-critical for L, and derive some consequences of (14.15).

Let

(14.16) $$R = \frac{\hbar}{2} \log \rho ,$$

so that $\nabla^i R = u^i$ and, by (14.15), $b^i = v^i + u^i = \nabla^i S - A^i + \nabla^i R$. If we write out (14.8) we obtain

$$\left(\frac{\partial}{\partial t} + b^i \nabla_i + \frac{\hbar}{2} \Delta\right) S = \frac{1}{2} b^i b_i + \frac{\hbar}{2} \nabla_i b^i - \phi + A_i b^i + \frac{\hbar}{2} \nabla_i A^i ,$$

and expressing everything in terms of R and S we find the *stochastic Hamilton-Jacobi equation*

(14.17) $$\frac{\partial S}{\partial t} + \frac{1}{2} (\nabla^i S - A^i)(\nabla_i S - A_i) + \phi - \frac{1}{2} \nabla^i R \nabla_i R - \frac{\hbar}{2} \Delta R = 0 .$$

(If $\hbar = 0$, then $R = 0$ and this reduces to the deterministic Hamilton-Jacobi equation.) We may rewrite (14.17), using (14.15), as

(14.18) $$\frac{\partial S}{\partial t} + \frac{1}{2} v^j v_j + \phi - \frac{1}{2} u^j u_j - \frac{\hbar}{2} \nabla^j u_j = 0 .$$

Now apply ∇_i and use (14.15) again. Then

$$(14.19)\quad \frac{\partial v_i}{\partial t} + \frac{\partial A_i}{\partial t} + \frac{1}{2}\nabla_i(v^j v_j) + \dot{\nabla}_i\phi - \frac{1}{2}\nabla_i(u^j u_j) - \frac{\hbar}{2}\nabla_i\nabla^j u_j = 0 .$$

Now u_j is a gradient, $u_j = \nabla_j R$, so its exterior derivative vanishes, and by (13.9) we have

$$(14.20)\quad \frac{1}{2}\nabla_i(u^j u_j) = u^j\nabla_j u_i .$$

By (14.15), v_j has the same exterior derivative as $-A_j$, so that

$$(14.21)\quad \frac{1}{2}\nabla_i(v^j v_j) = v^j\nabla_j v_i - v^j(\nabla_i A_j - \nabla_j A_i) = v^j\nabla_j v_i - H_{ij}v^j$$

(see (13.4)). By (10.10),

$$(14.22)\quad \frac{\hbar}{2}\nabla_i\nabla^j u_j = \frac{1}{2}\nabla_i\nabla^j\nabla_j R = \frac{1}{2}\nabla_i\Delta R = \frac{1}{2}\Delta\nabla_i R = \frac{1}{2}\Delta u_i .$$

Inserting (14.20), (14.21), and (14.22) into (14.19), we obtain

$$(14.23)\quad \frac{\partial v_i}{\partial t} + v^j\nabla_j v_i - u^j\nabla_j u_i - \frac{1}{2}\Delta u_i = -\nabla_i\phi - \frac{\partial A_i}{\partial t} + H_{ij}v^j .$$

But the left-hand side is the stochastic acceleration a^i of (10.13) (with the index i lowered) and the right-hand side is the force F_i (evaluated at the current velocity v^i), so that (14.23) is the *stochastic Newton equation*

$$(14.24)\quad m_{ij}a^j = F_i .$$

Just as in the deterministic case, a different choice of final time t_1 amounts merely to a gauge transformation that leaves the process ξ and (14.23) unchanged.

Following Yasue [14], let us seek the stochastic analogue of the first form of the principle of least action (in which the path is varied with fixed endpoints). Let ξ be a smooth diffusion on M, not necessarily Markovian. Let the Lagrangian L be as before, and define

$$(14.25)\quad J(\xi) = E\int_{t_0}^{t_1}\left[\tfrac{1}{2}L(\xi(t),D\xi(t),t) + \tfrac{1}{2}L(\xi(t),D_*\xi(t),t)\right]dt\,.$$

We say that ξ is *critical for* L *in the sense of Yasue* (Y-critical for L) in case it has finite energy and for all intervals $[t_0,t_1]$ and time dependent smooth vector fields X with compact support in $M\times(t_0,t_1)$,

$$(14.26)\qquad J(\xi+X) - J(\xi) = o(X)\,.$$

By Taylor's formula,

$$(14.27)\qquad J(\xi+X) - J(\xi) =$$

$$E\int_{t_0}^{t_1}\left[\frac{1}{2}\frac{\partial L}{\partial q^i}(\xi,D\xi,t)X^i + \frac{1}{2}\frac{\partial L}{\partial q^i}(\xi,D_*\xi,t)X^i + \frac{1}{2}\frac{\partial L}{\partial p^i}(\xi,D\xi,t)DX^i + \frac{1}{2}\frac{\partial L}{\partial p^i}(\xi,D_*\xi,t)D_*X^i\right]dt + o(X).$$

Recall that we are using D and D_* to denote stochastic derivatives with respect to $\mathcal{N}_t$ (not $\mathcal{P}_t$ and $\mathcal{F}_t$), and use the integration by parts formula (5.19). Since X vanishes at t_0 and t_1, (14.27) becomes

$$(14.28)\qquad J(\xi+X) - J(\xi) =$$

$$E\int_{t_0}^{t_1}\left[\frac{1}{2}\frac{\partial L}{\partial q^i}(\xi,D\xi,t) + \frac{1}{2}\frac{\partial L}{\partial q^i}(\xi,D_*\xi,t) - \frac{1}{2}D_*\frac{\partial L}{\partial p^i}(\xi,D\xi,t) - \frac{1}{2}D\frac{\partial L}{\partial p^i}(\xi,D_*\xi,t)\right]X^i dt + o(X)\,.$$

We may write the expression in square brackets as $B_i(\xi(t),t)$, where B_i is a smooth time dependent vector field. Then (14.26) is equivalent to the *stochastic Euler-Langrange equation*

(14.29) $$B_i = 0\,,$$

since we may take $X^i = f(t)B^i$ where f is any smooth function of compact support in (t_0,t_1). A straightforward computation shows that (14.29) is the same as (14.23) or (14.24). We have proved the following theorem.

THEOREM 14.2. *Let* L *be given by (14.1), where* ϕ *and* A_i *are smooth with compact support. Then a smooth diffusion* ξ *of finite energy is critical for* L *in the sense of Yasue if and only if the stochastic Euler-Lagrange equation (stochastic Newton equation) holds.*

Notice that the two action functionals J and I are not the same. The contributions from ϕ and A_i are the same, so consider the case of free motion (indicated by a subscript 0). Then

(14.30) $$J_0 = E\int_{t_0}^{t_1} \frac{1}{2}(v^jv_j + u^ju_j)dt\,,$$

but

(14.31) $$I_0 = E\int_{t_0}^{t_1} \frac{1}{2}(v^jv_j - u^ju_j)dt$$

since $\int_M \frac{\hbar}{2}(\nabla_j b^j)\rho = -\int_M b^ju_j\rho$ and $\frac{1}{2}b^jb_j - b^ju_j = \frac{1}{2}(v^jv_j - u^ju_j)$. Thus the osmotic energy $\frac{1}{2}u^ju_j$ is part of the kinetic energy in Yasue's formulation and is part of the potential energy in the Guerra-Morato formulation. We have seen that this is so, but I wish I understood why it is so.

Notice also that even if ξ is assumed to be Markovian, $\xi + X$ in general will not be. The two formulations are quite different conceptually, but both lead to the same equation (14.24).

Now let us consider the stochastic Hamilton-Jacobi equation (14.17) together with the current equation (5.24) expressed in terms of R and S :

$$(14.32)\quad \begin{cases} \dfrac{\partial S}{\partial t} + \dfrac{1}{2}(\nabla^i S - A^j)(\nabla_j S - A_j) + \phi - \dfrac{1}{2}\nabla^j R \nabla_j R - \dfrac{\hbar}{2}\Delta R = 0\,, \\[2ex] \dfrac{\partial R}{\partial t} + \nabla_j R(\nabla^j S - A^j) + \dfrac{\hbar}{2}\Delta S - \dfrac{\hbar}{2}\nabla_j A^j = 0\,. \end{cases}$$

Then (14.32) is a coupled system of nonlinear equations, but if we make the substitution

$$(14.33)\qquad \psi = e^{\frac{1}{\hbar}(R+iS)}$$

then a simple computation shows that (14.32) is equivalent to the *Schrödinger equation*

$$(14.34)\qquad i\hbar\frac{\partial\psi}{\partial t} = \left[\frac{1}{2}\left(\frac{\hbar}{i}\nabla^j - A^j\right)\left(\frac{\hbar}{i}\nabla_j - A_j\right) + \phi\right]\psi\,.$$

By a longer computation, one can derive the Schrödinger equation from the stochastic Newton equation together with the integrability condition that for some functions S, $v_i + A_i = \nabla_i S$.

The wave function ψ associated with a smooth diffusion ξ contains a wealth of information: it specifies the probability density $\rho = |\psi|^2$ and the drifts

$$(14.35)\qquad b^j_{\pm} = (\mathrm{Re} \pm \mathrm{Im})\hbar\,\nabla^j \log\psi \mp A^j$$

(it is sometimes convenient to write b_+ for b and b_- for b_*). If ξ is Markovian, then ψ determines ξ completely (with Neumann boundary conditions for the diffusion, if boundary conditions are relevant). If ξ is not Markovian, then ψ determines its *Markovian approximation*: the Markovian diffusion with the same diffusion tensor, probability density, and drifts as ξ.

Let $H(t)$ be the Hamiltonian operator in the right-hand side of (14.34). Then

$$(14.36)\qquad \langle\psi(t), H(t)\psi(t)\rangle = \int_M \left[\frac{1}{2}(u^j u_j + v^j v_j) + \phi\right]\rho = E\left[\frac{1}{2}(u^j u_j + v^j v_j) + \phi\right];$$

just the sig

lue of the *stoch*

ndent. Therefore E

ered the stronger conserva.

on); expressing $\mathscr{E}$ as a functio

$$\left(\frac{\partial}{\partial t} + v \cdot \nabla\right) \mathscr{E}(x,t) = 0.$$

chrödinger equation and diffusion pro-

yes [22]. In [47] I gave a derivation of this

newhat arbitrary booking form $\frac{1}{2}(DD_* + D_* D)\xi$

ation. Then Yasue [62] and Guerra-Morato [28]

on variational principles with stochastic Lagran-

uerra's result in §9, the derivation can be based on the

, if one is willing to accept the Pauli-DeWitt term.

[overlapping page fragment:] e solution of boundary condi- he Chapman- 1. To ∞ is ... sure e

§15. Nodes

simplicity of exposition (a standard euphemism for laziness on the

the author) I will assume in this section that M is compact. Then,

shown in the previous section, there is a bijective correspondence

between smooth Markovian diffusions with diffusion tensor $\hbar m^{ij}$ that are

critical for L and smooth nowhere zero solutions ψ of the Schrödinger

equation, the correspondence being determined by $\rho = |\psi|^2$ and (14.35).

Now let ψ be any smooth solution of the Schrödinger equation and let

Z be the nodes of ψ, i.e. the subset of $M \times R$ where ψ vanishes. Then

$b^j_\pm$ are well defined and smooth on Z^c, and can be used to define uniquely

a local Markovian diffusion there (see below), but the diffusing configura-

tion is in a quandary if it enters Z because the drifts are not defined there.

But it turns out that this problem is its own solution: the singularity of

the drifts on Z produces a repulsion strong enough to keep the configura-

tion from ever reaching the nodes.

Let $\varepsilon > 0$ and let

$$A_\varepsilon = \{(x,t) \in M \times R^+ : |\psi(x,t)| \leq \varepsilon\}$$

$0, \infty)$. For $0 \leq s < t$, let $p_\varepsilon(x,s;y,t)$ be th
Fokker-Planck equation on Z^c_ε with Dirichlet
initial value δ_x at time s. Then p_ε satisfies
gorov equation (6.5) but its integral (in y) is less tha
edy this, consider $\dot{M} = M \cup \{\infty\}$ where, since M is compa
n isolated point in $\dot{M}$. Define

$$p_\varepsilon(x,s;\{\infty\},t) = 1 - \int_M p_\varepsilon(x,s;y,t)\, d_M y \ ;$$

then p_ε is a transition probability. Choose as initial probability mea $\rho_{0\varepsilon} = \rho(0,y)\chi^c_\varepsilon(0,y)\, d_M y$, with $\rho_{0\varepsilon}(\{\infty\}) = 1 - \int \rho(0,y)\chi^c_\varepsilon(0,y)\, d_M y$, whe $\rho = |\psi|^2$ and χ^c_ε is the indicator function of Z^c_ε, and let $\Pr_\varepsilon$ be the corresponding regular probability measure on path space Ω (indexed by R^+). As usual, let $\xi(t)$ be the evaluation map $\omega \mapsto \omega(t)$; then ξ is a Markov process on $\dot{M}$. The configuration diffuses with drift b^j until it hits Z_ε, when it is killed (sent to ∞, where it rests eternally). The paths of ξ are continuous except at the moment of death.

Let ρ_ε be the probability density of ξ. Then $\rho_\varepsilon \leq \rho$, since both are solutions of the forward Fokker-Planck equation on Z^c_ε with the same initial value and $\rho_\varepsilon = 0$ on ∂Z^c_ε.

The ρ_ε are increasing in y on M as ε decreases. Let $p(x,s;y,t)$ be their limit, with $p(x,s;\{\infty\},t)$ the defect in its integral (we will show that this is 0), and let $\Pr$ be the corresponding regular probability measure on Ω with initial measure $\rho(0,y)\, d_M y$. Let

$$D = \{\omega : \omega(t) = \infty \text{ for some } t \text{ in } R^+\}.$$

Then $\Pr_\varepsilon(D)$ decreases to $\Pr(D)$.

THEOREM 15.1. $\Pr(D) = 0$.

Proof. Let $0 < T < \infty$ and let $D_T = \{\omega : \omega(t) = \infty \text{ for some } t \text{ in } [0,T]\}$.

Then we need only show that $\Pr_\varepsilon(D_T)$ decreases to 0. Throughout this proof, time parameters are restricted to lie in $[0,T]$.

Let us set $\hbar = 1$, and let R, as always, be defined by (14.16), so that $|\psi| = e^R$. Let

$$X(t) = R(\xi(t),t) - R(\xi(0),0) ,$$

with the convention that $R(\infty) = 0$, and let $X = \sup |X(t)|$. By the continuity of paths, D_T is equal $\Pr_\varepsilon$ a.e. to $\{\inf R(\xi(t),t) = \log \varepsilon\}$, so we need only establish bounds on $\Pr_\varepsilon\{X > \lambda\}$ that are independent of ε and tend to 0 as $\lambda \to \infty$.

Let w^i and w_*^i be defined (as difference processes) by $d\xi^i \equiv b^i dt + dw^i$ and $d_*\xi^i \equiv b_*^i dt + d_* w_*^i$. By (5.16),

$$X(t) = \int_0^t dR(\xi(s),s) = \int_0^t \left[\frac{\partial R}{\partial s}\, ds + b^j \nabla_j R\, ds + \nabla_j R\, \tilde{d}w^j(s) + \frac{1}{2}\Delta R\, ds\right]$$

$$= \int_0^t d_* R(\xi(s),s) = \int_0^t \left[\frac{\partial R}{\partial s}\, ds + b_*^j \nabla_j R\, ds + \nabla_j R \tilde{d}_* w_*^j(s) - \frac{1}{2}\Delta R\, ds\right] .$$

Recall that $\nabla_j R = u_j$ and $\frac{1}{2} b^j + \frac{1}{2} b_*^j = v^j$, and average these two expressions. Then

$$X(t) = \int_0^t \left[\frac{\partial R}{\partial s}\, ds + v^j u_j ds + \frac{1}{2} u_j \tilde{d} w^j(s) + \frac{1}{2} u_j \tilde{d}_* w_*^j(s)\right] .$$

Call the four integrals $X^\alpha(t)$ for $\alpha = 1,2,3,4$. Then $X^3(t) + X^4(t) = \int_0^t u_j \circ dw^j(s)$, but we will estimate them separately because X^3 is a martingale and X^4 is a martingale with time reversed. (An alternate way of estimating X, without using w_*^i, is given in [49].)

Let $\lambda > 0$. By Theorem 4.3,

$$\Pr_\varepsilon\{\sup |X^3(t)| > \lambda| \leq \frac{1}{4\lambda^2} \int_0^T \int_M u^j u_j \rho d_M x \, dt \, . \tag{15.1}$$

Let $H_o(t) = H(t) - \phi(t)$, so that by (14.36),

$$<\psi, H_o(t)\psi> = \frac{1}{2} \int_M (u^j u_j + v^j v_j) \rho \, d_M x \, . \tag{15.2}$$

Since ψ is smooth, the right-hand side of (15.1) is well defined, independent of ε, and tends to 0 as $\lambda \to \infty$.

We have the same estimate for X^4, and clearly

$$\Pr_\varepsilon\{X^2 > \lambda\} \leq \frac{1}{\lambda} \int_0^T \int_M |v^j u_j| d_M x \, dt \, ,$$

which is bounded in terms of (15.2) by the Schwarz inequality.

I would like to estimate X^1 similarly by an energy integral, but I don't see how to do this. However,

$$\Pr_\varepsilon\{X^1 > \lambda\} \leq \frac{1}{\lambda} \, E \int_0^T \left|\frac{\partial R}{\partial T}\right| dt = \frac{1}{\lambda} \int_0^T \int_{\{x : |\psi(x,t)| \geq \varepsilon\}} \frac{1}{2} \left|\frac{\partial \rho}{\partial t}\right| \rho^{-1} \rho_\varepsilon d_M x \, dt \tag{15.3}$$

$$\leq \frac{1}{\lambda} \int_0^T \int_M \frac{1}{2} \left|\frac{\partial \rho}{\partial t}\right| d_M x \, dt \, .$$

Now $\frac{\partial \rho}{\partial t} = i \, \overline{H(t)\psi(t)} \, \psi(t) - i \, \overline{\psi(t)} \, H(t)\psi(t)$, so that

$$\int_M \frac{1}{2} \left|\frac{\partial \rho}{\partial t}\right| d_M x \leq \|H(t)\psi(t)\|_2 \, .$$

Thus the right-hand side of (15.3) is well defined, independent of ε, and tends to 0 as $\lambda \to \infty$. The nodes are never reached. ■

We are interested in M that are not necessarily compact and ψ that are not necessarily smooth, so it is worth recording that we have an estimate on $\Pr\{X>\lambda\}$ that tends to 0 as $\lambda \to \infty$ in terms of

$$(15.4) \qquad \int_0^T \langle \psi(t), H_0\psi(t)\rangle\, dt \quad \text{and} \quad \int_0^T \|H(t)\psi(t)\|_2\, dt\,.$$

If H is time independent with the same domain as H_0, a situation of frequent occurrence in quantum mechanics, then (15.4) are finite if $\psi(0) \in \mathfrak{D}(H)$. Therefore this method should work in this general setting. I will not try to carry out the details since I expect that this method will soon be superseded, but we may proceed with the assurance that Markov processes associated with decent solutions of the Schrödinger equation exist, are unique (see also [67]), and do not ever reach the nodes.

Suppose that the nodes of ψ separate $M \times \mathbf{R}$ into disconnected regions G_i, with no communication among them. One's first inclination is to take $\rho = |\psi|^2$ as the probability density of the process. But let χ_i be the indicator function of G_i and let

$$p_i = \int_M \chi_i(x,t)\,\rho(x,t)\,d_M x\,.$$

These are independent of t, since there is no communication among the G_i, and $\Sigma p_i = 1$. Let $\rho_i = p_i^{-1}\chi_i\rho$, so that each ρ_i is a probability density. There is nothing to prevent us from taking $\Sigma \bar{p}_i\rho_i$ as the probability density for the diffusion, for any positive $\bar{p}_i$ summing to 1. From the point of view of conservative diffusion theory, the wave function ψ is just a convenience for computation, and no physical significance is attached to it.

However, if the $\bar{p}_i$ are not equal to the p_i, the resulting diffusion is not stable in a certain sense. Suppose the potentials are perturbed by a perturbation of order ε. Then the wave function ψ will be perturbed to a new wave function ψ_ε. Now generically the zeros of a complex function on a real manifold are of codimension two and do not disconnect the manifold, so for a generic perturbation, the nodes of ψ_ε will not disconnect $M \times \mathbf{R}$. Then the probability density $\rho_\varepsilon = |\psi_\varepsilon|^2$ is forced on us, but $\rho_\varepsilon \to \rho$. In exceptional circumstances when the nodes disconnect, we are free to alter the weighting in different regions – to choose $\bar{p}_i$ not equal to p_i –but the resulting process will not be close to the process obtained by a small perturbation of the potentials. (It might be objected that although the nodes of ψ_ε will not in general disconnect, the region where ρ_ε is small will disconnect, and we have established bounds showing that it is improbable to enter such a region in a given time interval. But the bounds we established were only logarithmic – to achieve a small probability of communicating among the G_i during a time interval of length T we would need to know the potentials to extreme accuracy, exponential in T.)

This discussion is qualitative, but it persuades me that the requirement that the diffusion have probability density $|\psi|^2$, where ψ satisfies the Schrödinger equation even at the nodes, is a necessary requirement in order to have the process stable under small perturbations of the potentials.

Chapter III
STOCHASTIC MECHANICS

In terms of the background field hypothesis, I see no reason to suppose that the diffusion of a system interacting with the background field is Markovian. But Markov processes are immensely simpler and easier to study, and according to the wise principle of first looking for a lost wallet under the lamplight, let us first study Markovian diffusions. By *Markovian stochastic mechanics* I mean the study of those Markovian diffusions that are critical for a basic Lagrangian. There is no ambiguity or leeway for making additional assumptions in this theory: once the configuration space and Lagrangian are decided on, we have a definite class of diffusions.

In quantum mechanics, when the wave function is known everything is known — that is the most complete description possible in quantum mechanics. But in stochastic mechanics that is just the starting point: we ask what the diffusion looks like. Whether this is an empty exercise or whether there is some physics involved will be discussed in Chapter IV.

§16. Gaussian Processes

It will be useful to have some examples in which the diffusion can be computed explicitly. In quantum mechanics, when the forces are linear and the wave function is (complex) Gaussian at one time, it remains Gaussian at all times. Then the associated diffusion will be (real) Gaussian, and to describe a Gaussian stochastic process completely we need only give its mean and covariance.

We take the configuration space M to be $\mathbf{R}^n$. I will usually set $\hbar = 1$ but on occasion I will insert factors of $\hbar$. We use the usual

Euclidean coordinates $x^1, \cdots, x^n$. The mass tensor m_{ij} is constant. If $n = 3N$ and we have a system of N particles of masses m_α, then m_{ij} will be a diagonal matrix with each m_α repeated three times. It will be useful though to retain our tensor notation. For example, the free Hamiltonian operator is $-\frac{\hbar}{2}\Delta$ with the masses automatically in the right place.

If we have a time dependent Hamiltonian operator $H(t)$, then the solution of the Schrodinger equation $\frac{\partial\psi}{\partial t} = -i\,H(t)\psi(t)$ satisfies $\psi(t) = U(s,t)\psi(s)$ for a unitary propagator U. In this section we will use Heisenberg operators: $X^j(0)$ is multiplication by x^j, $P_j(0)$ is $-i\frac{\partial}{\partial x^j}$, $X^j(t) = U(t,0)X^j(0)U(0,t)$, and $P_j(t) = U(t,0)P_j(0)U(0,t)$. For any operator A we let $\langle A\rangle = \langle\psi_0, A\psi_0\rangle$, and we abbreviate $\frac{1}{2}(AB+BA)$ by $A \circ B$.

In this section we assume that

$$H(t) = \frac{1}{2}(-i\nabla^j - A^j(t))(-i\nabla_j - A_j(t)) + \phi(t)$$

where for each t, $\phi(t)$ is a polynomial in x of degree at most two and $A_j(t)$ is a polynomial in x of degree at most one. Then the forces are linear, and to find the Heisenberg position and velocity operators $X^j(t)$ and $P^j(t)$ explicitly in terms of their initial values $X^j(0)$ and $P^j(0)$, we need to solve precisely the same set of linear differential equations as in deterministic mechanics. In simple cases this can be done in closed form or by quadratures, and in the general case there is an explicit representation in terms of product integrals.

Assuming the $X^j(t)$ and $P^j(t)$ known, we can find an explicit expression for $\psi(t)$ if $\psi(0)$ is a given Gaussian. The explicit computation in simple looking cases is more complicated than one might think. Perhaps this elementary topic is systematized somewhere in the literature, but here is one way of doing it.

Write the complex Gaussian ψ in the form e^{R+iS} where

(16.1) $$R = -\frac{1}{4}\,\sigma_{jk}^{-1}(x^j-\mu^j)(x^k-\mu^k)+\gamma\,,$$

(16.2) $$S = \frac{1}{2}\,\tau_{jk}(x^j-\mu^j)(x^k-\mu^k)+\nu_j x^j\,.$$

All these quantities are real, σ_{jk}^{-1} and τ_{jk} are symmetric, and σ_{jk}^{-1} is of strictly positive type; all are time dependent. The γ is a normalization constant chosen to make $\|\psi\|_2 = 1$. We can express the other quantities in terms of the Heisenberg operators. Let σ^{jk} be the inverse matrix, $\sigma^{jk}\sigma_{k\ell}^{-1} = \delta_\ell^j$. Then

(16.3) $$\mu^j = \langle X^j\rangle\,,$$

(16.4) $$\nu_j = \langle P_j\rangle\,,$$

(16.5) $$\sigma^{jk} = \langle (X^j-\mu^j)(X^k-\mu^k)\rangle\,,$$

(16.6) $$\tau_{jk} = \sigma_{ja}^{-1}\langle (X^a-\mu^a)\circ P_k\rangle\,.$$

It suffices to verify these formulas at $t = 0$. Now $|\psi|^2 = e^{2R}$ is a Gaussian probability density, and (16.3) and (16.5) are familiar to every probabilist: the mean is μ^j and the covariance is σ^{jk}. Since $\langle X^j-\mu^j\rangle = 0$, the only contribution to $\left(\psi(0), -i\,\frac{\partial}{\partial x^j}\,\psi(0)\right)$ is from the linear term in S, so (16.4) holds. We have, at $t = 0$,

(16.7) $$\langle (X^a-\mu^a)\circ P_j\rangle = \left\langle \psi(0), \left[(x^a-\mu^a)\,\frac{1}{i}\,\frac{\partial}{\partial x^j}+\frac{1}{2i}\,\delta_j^a\right]\psi(0)\right\rangle.$$

The ν_j term contributes nothing, and the σ^{-1} term cancels the $\frac{1}{2i}\,\delta_j^a$. Therefore (16.7) is $\sigma^{ak}\tau_{kj}$, so (16.6) holds. Let me also record the fact that

(16.8) $$\langle (P_j-\nu_j)(P_k-\nu_k)\rangle = \frac{1}{4}\,\sigma_{jk}^{1}+\tau_{ja}\sigma^{ab}\tau_{bk}\,.$$

All of this is just a recipe for solving the Schrödinger equation in the simples possible case. Now let us look at the diffusion. By (16.1) and (16.2),

$$u_j = -\frac{1}{2}\sigma^{-1}_{jk}(x^k - \mu^k)\,, \tag{16.9}$$

$$v_j = \tau_{jk}(x^k - \mu^k) + \nu_j\,. \tag{16.10}$$

Now $d\xi_j = (u_j + v_j)dt + dw_j$, where w is the Wiener process, and since $d\mu_j/dt = \nu_j$ we have the linear stochastic differential equation

$$d(\xi_j - \mu_j) = a_j^k(\xi_k - \mu_k)\,dt + dw_j(t) \tag{16.11}$$

where

$$a_j^k = -\frac{1}{2}\,\sigma^{-1}{}_j{}^k + \tau_j{}^k\,. \tag{16.12}$$

This is a time dependent $n \times n$ matrix, and (16.11) has an explicit solution in terms of product integrals, which I write as time-ordered exponentials:

$$(\xi_j - \mu_j)(t) = \mathcal{T}e^{\int_s^t a_j^k(r)dr}(\xi_k - \mu_k)(s) + \int_s^t \mathcal{T}e^{\int_r^t a_j^k(z)dz}\,dw_k(r)\,. \tag{16.13}$$

Then $\xi^j - \mu^j$ is the Gaussian process with mean 0 and covariance

$$E(\xi^j(t) - \mu^j(t))(\xi^k(s) - \mu^k(s)) = \mathcal{T}e^{\int_s^t a^j_a(r)dr}\,\sigma^{ak}(s)\,, \quad s \le t\,. \tag{16.14}$$

Here is the simplest example, which I call the *one-slit process*. Take $\mu^j(0), \nu_j(0)$, and $\tau_{jk}(0)$ to be 0, and $\sigma^{jk}(0) = \lambda^2\delta^{jk}$. (Think of a particle traveling through a slit – a "Gaussian slit" for ease of computation – of half-width λ at time 0, in a frame of reference comoving with the beam.) The particle is free, so $X^j(t) = X^j(0) + tP^j(0)$ and $P^j(t) = P^j(0)$. Then $\mu^j(t) = 0$, $\nu_j(t) = 0$, and using (16.8) we find

$$(16.15)\qquad \sigma^{jk}(t) = \langle (X^j(0)+tP^j(0))(X^k(0)+tP^k(0))\rangle = $$

$$= \left(\lambda^2 + \frac{t^2}{4\lambda^2}\right)\delta^{jk} = \frac{4\lambda^4+t^2}{4\lambda^2}\,\delta^{jk}\,,$$

$$(16.16)\qquad \tau_{jk}(t) = \sigma^{-1}_{ja}(t)\langle (X^a(0)+tP^a(0))\circ P_k(0)\rangle = \frac{t}{4\lambda^4+t^2}\,\delta_{jk}\,,$$

$$(16.17)\qquad u^j = -\frac{2\lambda^2}{4\lambda^4+t^2}\,x^j\,,$$

$$(16.18)\qquad v^j = -\frac{t}{4\lambda^4+t^2}\,x^j\,,$$

$$(16.19)\qquad E\,\frac{1}{2}\,u^ju_j = \frac{1}{8\lambda^2}\,\frac{4\lambda^4}{4\lambda^4+t^2}\,,$$

$$(16.20)\qquad E\,\frac{1}{2}\,v^jv_j = \frac{1}{8\lambda^2}\,\frac{t^2}{4\lambda^4+t^2}\,.$$

The expected kinetic energy $E\,\frac{1}{2}\,(u^ju_j+v^jv_j)$ is a constant, $1/8\lambda^2$. At $t=0$ it is all osmotic energy, but as time goes on the energy is transformed into current energy. We have

$$(16.21)\qquad \alpha^k{}_j(t) = \frac{t-2\lambda^2}{4\lambda^4+t^2}\,\delta^k_j\,,$$

so ξ^j is Gaussian of mean 0 and covariance

$$(16.22)\quad E\xi^j(t)\xi^k(s) = \frac{1}{4\lambda^2}\sqrt{4\lambda^4+t^2}\sqrt{4\lambda^4+s^2}\,e^{-\arctan\frac{t}{2\lambda^2}+\arctan\frac{s}{2\lambda^2}}\,\delta^{jk},\quad s\le t\,.$$

By (16.22), $E\left(\frac{\xi^j(s)}{s}-\frac{\xi^j(t)}{t}\right)\left(\frac{\xi_j(s)}{s}-\frac{\xi_j(t)}{t}\right)\to 0$ as $s,\ t\to\infty$. That is, the limit average velocity

(16.23)
$$\pi_+^j = \lim_{t\to\infty} \frac{\xi^j(t)}{t} = \lim_{t\to\infty} \frac{\xi^j(t)-\xi^j(0)}{t}$$

exists in L^2 (and therefore a.e. on Ω since the random variables are Gaussian). The limit momentum π_{+j} is Gaussian of mean 0 and covariance $\frac{1}{4\lambda^2}\delta_{jk}$, so its probability density is the square modulus of the Fourier transform of ψ. This is a very special case of a general result of Shucker (see §18). We have $\xi^j(t) = t\pi_+^j + o(t)$ a.e. on Ω as $t \to \infty$, so a.e. on Ω, $u^j(\xi(t),t) \to 0$ and $v^j(\xi(t),t) \to \pi_+^j$. Asymptotically, the particle travels in a straight line with velocity π_+^j, subject to the random fluctuations of a superimposed Wiener process.

In the same way,

$$\pi_-^j = \lim_{t\to-\infty} \frac{\xi^j(t)}{t}$$

exists, and using (16.22) we find that the correlation coefficient of π_+^j and π_-^k is δ^{jk} times $-e^{-\pi} = -.04321\cdots$. This curious correlation, not depending on λ, is mentioned to point out that if anyone undertakes to study scattering theory in stochastic mechanics, the framework will have to be formulated carefully.

The wave function for a beam of particles of momentum ν_j is

(16.24)
$$\psi = e^{i\nu_j x^j - i\nu_j \nu^j t}.$$

This has $u^j = 0$, $v^j = \nu^j$, but $\rho = 1$; it is nonnormalizable. One can easily make sense of a nonnormalizable diffusion: use the same formula to construct the measure on path space but with an infinite initial measure (one even obtains a regular measure on the locally compact space $\Omega \setminus \{\omega : \omega(0) = \infty\}$). For example, one has the two-sided Wiener process with $\rho = 1$, $u^j = 0$, $v^j = 0$ (corresponding to the solution $\psi = 1$ of the free Schrödinger equation). The diffusion associated with (16.24) is simply the sum of motion with constant velocity ν^j and the two-sided Wiener process.

Now let us discuss the harmonic oscillator. We take $n = 1$ and the mass $m = 1$, and we denote the circular frequency by ω. Then

$$\text{(16.25)} \qquad \begin{cases} X(t) = \cos\omega t\, X(0) + \omega^{-1}\sin\omega t\, P(0) \\ P(t) = -\omega\sin\omega t\, X(0) + \cos\omega t\, P(0)\,. \end{cases}$$

Consider first the ground process

$$\text{(16.26)} \qquad \psi = \left(\frac{\omega}{\pi}\right)^{\frac{1}{4}} e^{-\frac{\omega}{2}x^2}\,.$$

This is a stationary process with $u = -\omega x$, $v = 0$, and $\rho = \psi^2$. Then ξ is Gaussian with mean 0 and covariance

$$\text{(16.27)} \qquad E\xi(t)\xi(s) = \frac{1}{2\omega}\, e^{-\omega|t-s|}\,.$$

This is the only stationary Gaussian Markov process, and it is not surprising that it comes up in many different contexts. It is familiar in the theory of dissipative diffusion as the velocity process in Ornstein-Uhlenbeck theory, and in constructive field theory as the free Euclidean (imaginary time) field in one dimension (zero space dimensions). But here it occurs in the real time theory of conservative diffusion. By (16.27) there is a relaxation time ω^{-1}, and the quantum fluctuations destroy correlation in positions of the particle at two widely separated times. Because of the osmotic velocity $u = -\omega x$, there is a constant tendency for the particle to move towards 0, and the strength of this tendency is directly proportional to displacement from 0. As always with osmotic effects, this is symmetric under time reversal. If one takes a motion picture of ξ, it looks as if it is constantly trying to get back to 0, and the movie looks just the same if it is run backwards.

This process satisfies the stochastic Newton equation $\frac{1}{2}(DD_* + D_*D)\xi(t) = -\omega^2\xi(t)$. By the linearity of this equation for the harmonic oscillator, if we add a solution of the deterministic equation $\frac{d^2}{dt^2}\mu(t) = -\omega^2\mu(t)$, we get a solution again. The new process $\eta(t) = \xi(t) + \mu(t)$ is the deterministic motion with the random fluctuations of the ground process superimposed. This is the coherent process with wave function

$$(16.28) \qquad \left(\frac{\omega}{\pi}\right)^{\frac{1}{4}} e^{-\frac{\omega}{2}(y-\mu(t))^2 + ix\nu(t) - \frac{i}{2}\nu(t)\mu(t) - \frac{i}{2}\omega t}$$

where $\frac{d\mu}{dt} = \nu(t)$. This elegant fact was discovered by Guerra and Loffredo [74] and has been exploited by Ruggiero and Zannetti [54][55][56][70]. For a comprehensive review of coherent states, see [51].

§17. Interference

The Schrödinger equation is linear, so if ψ_1 and ψ_2 are solutions, then $\psi = \psi_1 + \psi_2$ is also. Let $\psi_1 = e^{R_1+iS_1}$, $\psi_2 = e^{R_2+iS_2}$, and $\psi = e^{R+iS}$ on the complement of their nodes, and let u_1, u_2, u, v_1, v_2, v be the gradients of R_1, R_2, R, S_1, S_2, S (though it must be remembered that in the presence of covector potentials, the v's are not current momenta). Then, on the complement of the nodes,

$$(17.1) \quad u = \frac{1}{2}(u_1+u_2) + \frac{1}{2}\,\frac{\sinh(R_1-R_2)(u_1-u_2) - \sin(S_1-S_2)(v_1-v_2)}{\cosh(R_1-R_2) + \cos(S_1-S_2)},$$

$$(17.2) \quad v = \frac{1}{2}(v_1+v_2) + \frac{1}{2}\,\frac{\sinh(R_1-R_2)(v_1-v_2) + \sin(S_1-S_2)(u_1-u_2)}{\cosh(R_1-R_2) + \cos(S_1-S_2)}.$$

To see this, multiply the equation $e^{R_1+iS_1} + e^{R_2+iS_2} = e^{R+iS}$ by its complex conjugate and factor out $e^{R_1+R_2}$ on the left-hand side, obtaining

$$(17.3) \qquad 2\,e^{R_1+R_2}\left[\cosh(R_1-R_2) + \cos(S_1-S_2)\right] = e^{2R}.$$

Take logarithmic gradients to obtain (17.1). Now write ψ_1 in Cartesian form $\psi_1 = X_1 + iY_1$, so that $X_1 = e^{R_1}\cos S_1$ and $Y_1 = e^{R_1}\sin S_1$, and similarly for ψ_2 and ψ. Then $S = \arctan(Y/X)$, so that $v = e^{-2R}[X\nabla Y - Y\nabla X]$. Perform the differentiations and use (17.3) and the addition formulas for sin and cos to obtain (17.2).

If one sends a beam of electrons through a crystal, one observes a diffraction pattern. For purposes of discussion, the crystal is frequently replaced by a screen with two slits in it. Let the slits be Gaussian, of half-width λ, located at $\pm a$, where a is a 3-vector. We set $m = 1$ and $\hbar = 1$, and use a frame of reference comoving with the beam. Then the wave function is

$$\psi(x,t) = \gamma(\psi_1(x-a,t) + \psi_1(x+a,t)) \tag{17.4}$$

where ψ_1 is the wave function of the one-slit process discussed in the previous section. Using (17.1), (17.2), and the formulas of the previous section we may calculate the drift b. The result of the computation, using vector notation, is

$$b = \frac{t-2\lambda^2}{4\lambda^4+t^2}\,x + \frac{\left(\sinh\dfrac{4\lambda^2 a\cdot x}{4\lambda^4+t^2}\right)\dfrac{t-2\lambda^2}{4\lambda^4+t^2} - \left(\sin\dfrac{2ta\cdot x}{4\lambda^4+t^2}\right)\dfrac{t+2\lambda^2}{4\lambda^4+t^2}}{\cosh\dfrac{4\lambda^2 a\cdot x}{4\lambda^4+t^2} + \cos\dfrac{2ta\cdot x}{4\lambda^4+t^2}}\,a\,. \tag{17.5}$$

Only the direction joining the two slits is of interest, so let us fix a to be the unit vector on the x-axis and treat x as a coordinate rather than a vector, so that $a\cdot x = x$. Then λ is a dimensionless parameter less than unity, the ratio of the slit width to the separation of the two slits.

This process has no nodes, because $\cosh > 1$ unless $x = 0$, in which case $\cos = 1$. However, it comes very close to having nodes when

$$\frac{2tx}{4\lambda^4+t^2} = (2n+1)\pi \tag{17.6}$$

and $4\lambda^2 x/(4\lambda^4+t^2)$ is small. The characteristic time for this process is λ (or, keeping λ dimensionless and reinserting the dimensional factors, $\lambda ma^2/\hbar$). For times less than λ, the last term in (17.5) is small. The equation (17.6) is a hyperbola, but for times bigger than λ it is practically its asymptote $x=(2n+1)\frac{\pi}{2}t$. When t is an order of magnitude greater than λ, the drift is enormous near the hyperbolas, pointing away from them. The situation is illustrated in Figure 1 for $\lambda=\frac{1}{10}$.

Before time λ, the particle, entering from one of the slits, diffuses very much as a particle in the one-slit process, but soon after time λ it finds itself trapped in one of the channels.

We may condition the process to enter the top slit: let $\hat{\rho}_+(x,0)=\psi_1(x-1,0)^2$. We cannot compute the conditioned probability density $\hat{\rho}_+(x,t)$ exactly, because there is no Schrödinger equation associated to it, and for times bigger than λ it will exhibit a diffraction pattern skewed upward. We may also condition the process to enter the bottom slit, with $\hat{\rho}_-(x,0)=\psi_1(x+1,0)^2$. Since $\psi_1(x-1,0)$ and $\psi_1(x+1,0)$ are almost orthogonal, $\rho(x,0)$ is almost the average of $\hat{\rho}_+(x,0)$ and $\hat{\rho}_-(x,0)$. In fact,

$$\rho(x,0)=\left(\frac{1}{2}-\frac{\delta}{2}\right)\hat{\rho}_+(x,0)+\left(\frac{1}{2}-\frac{\delta}{2}\right)\hat{\rho}_-(x,0)+\delta\hat{\rho}_0(x,0) \tag{17.7}$$

where $\hat{\rho}_0(x,0)=(2\pi\lambda^2)^{-1/2}e^{-x^2/2\lambda^2}$ and δ is the very small number

$$\delta=\frac{1}{2}-\frac{1}{2}\left(1+e^{-\frac{1}{2\lambda}}\right)^{-1}$$

which measures the overlap of the two Gaussian slits. Of course, (17.7) continues to hold for all times t; to within δ, $\rho(x,t)$ is the average of $\hat{\rho}_+(x,t)$ and $\hat{\rho}_-(x,t)$. It would be fun to see a computer-generated movie of this process, with particles diffusing from the slits and accumulating on another screen at time 10λ.

The interpretation presented here flatly contradicts the customary discussion of interference (see e.g. [24, §1-1]). We find that (I) each particle

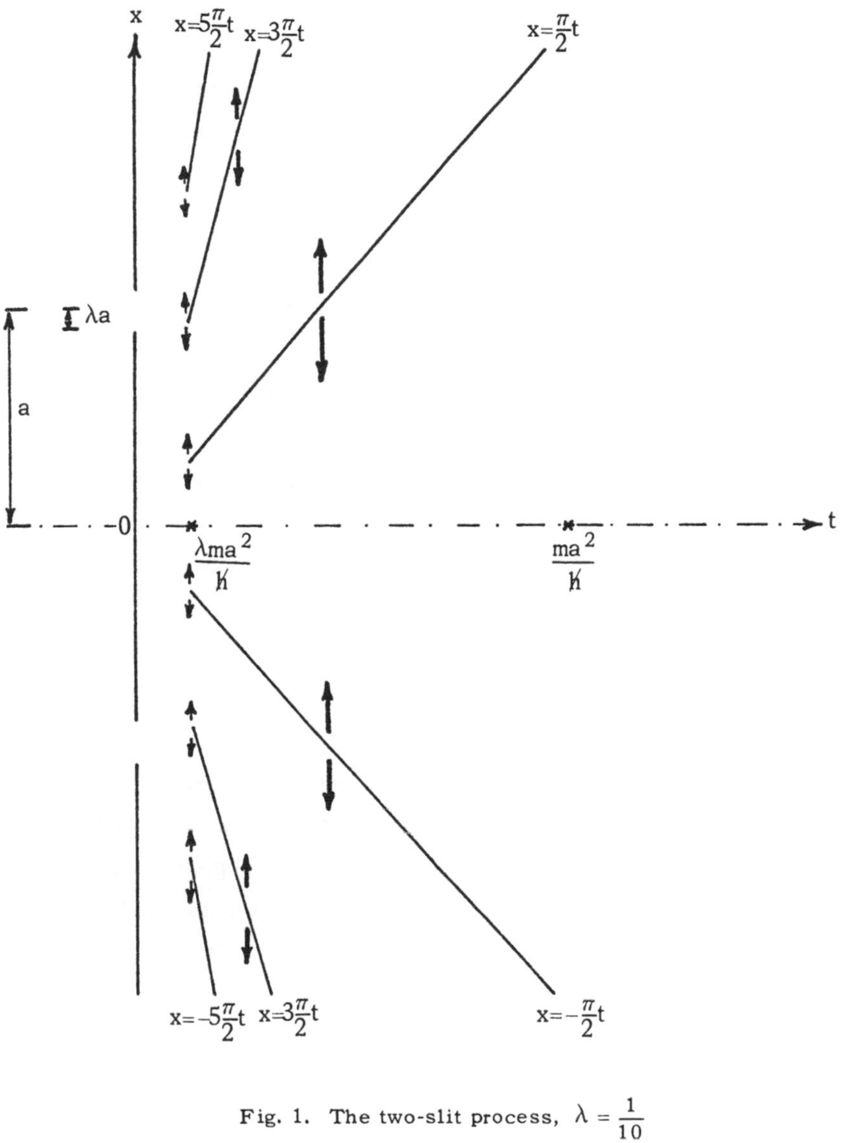

Fig. 1. The two-slit process, $\lambda = \frac{1}{10}$

must go through either the top slit or the bottom slit, and (II) the probability of arrival at a given point is the sum of two parts, the probability ρ_+ of arrival coming through the top slit and the probability ρ_- of arrival coming through the bottom slit. The denial of (II) in [24] is the consequence of confounding the conditioned two-slit process with the one-slit process.

The title of this section is a misnomer: nothing is interfering with anything.

Nevertheless, one may reasonably ask: why doesn't the particle, entering the top slit, diffuse according to the one-slit process? The probability density $\rho(x,0)$ is prepared by a physical interaction: sending the beam of particles to the screen. Then the only Markovian diffusion with the particles localized with equal probability at the two slits, with zero current velocity initially (in our frame of reference), with diffusion constant $\hbar/m$, and with zero stochastic acceleration, is the one we have discussed. Similarly, if the particle is localized at the upper slit by a physical interaction (e.g. closing the lower slit), it will diffuse according to the one-slit process. Conditioning is not a physical interaction; it is a mathematical device for constructing a measure on path space. Its interpretation is that if we could look at the particles without disturbing their motion (no way of doing this is yet known, but the consistency of the theory does not rule out the possibility—we will return to this point in §22), the conditioned process describes what we would see for those particles happening to start from the upper slit.

Still one may ask: how does the particle, entering the top slit, know that the bottom slit is open? According to the background field hypothesis, the diffusion is the result of an interaction with the background field. The electromagnetic field is different in the presence of a crystal, and the electrons accordingly have different random fluctuations due to interaction with it. The particle doesn't know whether the other slit is open, but the background field does. One might attempt to refute this picture by observing the probability density at a time less than the time $2a/c$ it takes light to travel from one slit to the other. But to have an appreciable difference between the one-slit and two-slit processes we would need

$$\frac{2a}{c} \gtrsim \frac{\lambda m a^2}{\hbar}, \quad \text{or} \quad \lambda a \lesssim 2\frac{\hbar}{mc};$$

i.e. the slit would have to be so narrow as to localize the particle to within two Compton wavelengths. This certainly cannot be done with a crystal, and doing it by any means would produce a disturbance in the momentum of relativistic magnitude — but then the basis for discussing the problem in terms of the nonrelativistic Schrödinger equation would be destroyed.

§18. Momentum

Let ψ be a solution of the free Schrödinger equation on $\mathbf{R}^n$ that lies in the Schwartz space $\mathcal{S}(\mathbf{R}^n)$ (for one time, and therefore for all times). The technique of §15 certainly applies in this case (or see [67]), and there is a unique Markov process ξ associated with ψ. Let $\hbar = 1$ and $m = 1$, and let $\pi(t) = \xi(t)/t$. Then $|\psi(x,t)|^2$ is the probability density of $\xi(t)$, so $|\psi(pt,t)|^2$ is the probability density of $\pi(t)$. A simple calculation shows that

$$\lim_{t\to\infty} |\psi(pt,t)|^2 t^n = |\tilde{\psi}(p)|^2 \tag{18.1}$$

where $\tilde{\psi}$ is the Fourier transform of ψ. The way one usually measures the momentum of a particle is to measure $\pi(t)$ for a large t (in a coordinate system for which it is known that $\xi(0)$ is approximately 0). Knowing that the probability density of $\pi(t)$ has a limit says nothing about the behavior of the individual paths, but Shucker [57] has proved the following theorem:

THEOREM 18.1. *$\lim_{t\to\infty} \pi(t)$ exists a.e. on Ω.*

Proof. Formulas labeled with a prefix S correspond to the formulas in [57], though sometimes with a change in notation. To begin with,

$$\nabla\psi(x,t) = \frac{ix}{t}\psi(x,t) - \int \frac{iy}{t} e^{-n\pi i/4}(2\pi t)^{-n/2} e^{i(x-y)^2/2t}\psi(y,0)\,dy\,. \tag{S27}$$

Now $b(x,t) = (\mathrm{Re} + \mathrm{Im})\nabla\psi/\psi$, so

(S28) $$b(x,t) - \frac{x}{t} = 0(t^{-1-\frac{n}{2}}|\psi(x,t)|^{-1}),$$

(S29) $$\begin{aligned} d\xi(t) &= b(\xi(t),t)\,dt + dw(t) \\ &= \frac{\xi(t)}{t}\,dt + 0(t^{-1-\frac{n}{2}}|\psi(\xi(t),t)|^{-1})\,dt + dw(t)\,, \end{aligned}$$

(S31) $$d\rho(t) = \frac{d\xi(t)}{t} - \frac{\xi(t)}{t^2}\,dt = 0(t^{-2-\frac{n}{2}}|\psi(t\pi(t),t)|^{-1})\,dt + \frac{1}{t}\,dw(t)\,.$$

Now define

(S32) $$A_{\delta_1} = \{p : |\tilde{\psi}(p)| > \delta_1\}\,,$$

(S33) $$A_{\delta_1,\delta_2} = \{p : \text{ if } |p'-p| < \delta_2 \text{ then } p' \in A_{\delta_1}\}\,.$$

Given $\varepsilon > 0$, we may choose δ_1 so that

(S34) $$\int_{A_{\delta_1}} |\tilde{\psi}(p)|^2\,dp > 1 - \frac{\varepsilon}{8}\,.$$

Now $\tilde{\psi}$ is continuous, so A_{δ_1} is open, and for δ_2 sufficiently small,

(S35) $$\int_{A_{\delta_1,\delta_2}} |\tilde{\psi}(p)|^2\,dp > 1 - \frac{\varepsilon}{4}\,.$$

By (18.1), we may choose T so that for all $t \geq T$,

(S36) $$\left||\psi(pt,t)|\,t^{\frac{n}{2}} - |\tilde{\psi}(p)|\right| < \frac{\delta_1}{2}\,.$$

For $t \geq T$ and $\pi(t)$ in A_{δ_1}, we then have

(S37) $$|\psi(pt,t)|\, t^{\frac{n}{2}} > \frac{\delta_1}{2} .$$

By (S31), when $\pi(t) \in A_{\delta_1}$,

(S38) $$d\pi(t) = 0(t^{-2})\,dt + \frac{1}{t}\,dw .$$

Choose δ so that $0 < \delta < 2\delta_2$. Increase T if need be so that

(S39) $$\int_T^\infty |0(t^{-2})|dt < \frac{\delta}{4} ,$$

(S40) $$\Pr\left\{\sup_{s>T} \left| \int_T^s \frac{1}{t}\, dw(t)\right| > \frac{\delta}{4}\right\} < \frac{\varepsilon}{2}$$ (this is possible by Theorem 4.2),

(S41) $$\int_{A_{\delta_1,\delta_2}} |\psi(pT,T)|^2\, T^n\, dp > 1 - \frac{\varepsilon}{2} .$$

Then if $\pi(T) \in A_{\delta_1,\delta_2}$ we have

(S42) $$\Pr_T\left\{\sup_{s>T} |\pi(s) - \pi(T)| > \frac{\delta}{2}\right\} < \frac{\varepsilon}{2} ,$$

where $\Pr_T$ is the conditional probability given $\mathfrak{N}_T$. But if $\pi(T) \in A_{\delta_1,\delta_2}$ and $|\pi(s) - \pi(T)| < \frac{\delta}{2}$, then $\pi(s) \in A_{\delta_1}$. Let

$$\Lambda_{\varepsilon,\delta} = \Omega \setminus \{\omega : \pi(T) \in A_{\delta_1,\delta_2} \text{ and } \pi(s) \in A_{\delta_1} \text{ for all } s \geq T\} .$$

By (S41) and (S42), $\Pr \Lambda_{\varepsilon,\delta} < \varepsilon$. But

$$\{\omega : \sup_{s_1,s_2 \geq T} |\pi(s_1) - \pi(s_2)| > \delta\} \subseteq \Lambda_{\varepsilon,\delta} ,$$

so

$$\Pr\{\omega: \sup_{s_1,s_2\ge T} |\pi(s_1)-\pi(s_2)| > \delta\} \le \varepsilon ,$$

$$\Pr\{\omega: \inf_T \sup_{s_1,s_2\ge T} |\pi(s_1)-\pi(s_2)| > \delta\} = 0 ,$$

and since this is true for all $\delta > 0$, the theorem is proved. ∎

Let $\pi_+ = \lim_{t\to\infty} \pi(t)$. The probability density of π_+ is $|\tilde{\psi}(p)|^2$, and asymptotically $\xi(t)$ moves in a straight line with velocity π_+ with the random fluctuations of the Wiener process superimposed.

§19. Bound States

Let H be a self-adjoint operator $H = H_0 + V$ on $L^2(R^n)$ where $H_0 = -\frac{1}{2}\Delta$ and V is a time independent multiplication operator, and suppose that H has an isolated eigenvalue $\lambda_0 = \inf \sigma(H)$, where $\sigma(H)$ is the spectrum of H, with eigenfunction $\psi_0 > 0$. Let $\rho = \psi_0{}^2$. Then ψ_0 (i.e. multiplication by ψ_0) is a unitary operator from $L^2(\rho dx)$ to $L^2(dx)$, and under this map the operator $\frac{1}{2}\Delta_\rho = \frac{1}{2}\Delta + u^j\nabla_j$ of §11 is unitarily equivalent to $\lambda_0 - H$ on $L^2(dx)$. Let $P^t = e^{\frac{t}{2}\Delta_\rho}$ on $L^2(\rho\, dx)$; then P^t is unitarily equivalent to $e^{(\lambda_0 - H)t}$ on $L^2(dx)$ (see [1]). Thus the same diffusion process may be looked at in two different but mathematically equivalent ways: as a real time conservative diffusion and as a process obtained by analytic continuation of the propagator for the Schrödinger equation to imaginary time. It is my hunch that the first point of view is more fundamental, and that it will lead to new methods of applying probability theory to quantum field theory. Analytic continuation in time may have obscured the issue of what Euclidean field theory is all about.

But for the present let us stay with simpler systems. Let μ be a probability measure with a density f with respect to ρdx, and suppose f to be in $L^2(\rho dx)$. Then $f = 1 + f_o$ where f_o is orthogonal to 1. Let $\lambda_1 = \inf(\sigma(H) \setminus \{\lambda_0\})$. Then

$$\|P^t f_o\|_{2,\rho} \leq e^{-t(\lambda_1 - \lambda_o)} \|f_o\|_{2,\rho} . \tag{19.1}$$

Let us call $(\lambda_1 - \lambda_o)^{-1}$ the *relaxation time*. It gives the time scale on which the quantum fluctuations produce decay of correlations between positions of ξ at different times.

Consider the hydrogen atom. Then $n = 3$, and we set $m = \hbar = e = 1$, so that the Schrödinger equation for stationary states of the hydrogen atom is

$$\left(-\frac{1}{2}\Delta - \frac{1}{r}\right)\psi = \lambda\psi , \tag{19.2}$$

where $r = |x|$ is the Euclidean norm of the vector x. The eigenvalue λ is the expectation of the stochastic energy. In the following discussion, γ denotes a normalization constant, possibly different in different cases.

(i) The wave function in the ground state is $\psi_o = \gamma e^{-r}$, $\lambda_o = -\frac{1}{2}$. Then u is the unit vector $u = -x/r$, and $v = 0$. A movie of this process, run forward or backward, shows a constant tendency to head towards the origin. Exercise: show that $\xi(t)$ never reaches the origin.

(ii) Let $\psi_1 = \gamma e^{-\frac{r}{2}}\left(1 - \frac{1}{2}r\right)$, $\lambda_1 = -\frac{1}{8}$. Here $R = \log\gamma - \frac{r}{2} + \log\left|1 - \frac{1}{2}r\right|$ and $u = \frac{-4+r}{4-2r}\frac{x}{r}$, $v = 0$. The nodal set is the sphere $r = 2$, and the singularity of u prevents $\xi(t)$ from reaching it. For $r > 4$, u points inward and prevents $\xi(t)$ from wandering off.

In this example, the nodal surface divides the configuration space into two disconnected regions with no communication between them. See the discussion at the end of §15.

(iii) Let $\psi_2 = \gamma e^{-\frac{r}{2}} r \cos\theta$, $\lambda_2 = -\frac{1}{8}$. Here we are using spherical coordinates ($x^1 = r \sin\theta\cos\phi$, $x^2 = r\sin\theta\sin\phi$, $x^3 = r\cos\theta$). Then $R = \log\gamma - \frac{r}{2} + \log r + \log|\cos\phi|$ and

$$u\cdot\nabla = \left(-\frac{1}{2}+\frac{1}{r}\right)\frac{\partial}{\partial r} + \tan\theta\,\frac{\partial}{\partial\theta}.$$

The nodal set is the plane $x^3 = 0$ $\left(\theta = \frac{\pi}{2}\right)$. The drift always points towards the sphere $r = 2$ and away from the nodal plane, with a tendency to move toward the x^3-axis.

(iv) Let $\psi_3 = \gamma e^{-\frac{r}{2}} r \sin\theta\, e^{i\phi}$, $\lambda_3 = -\frac{1}{8}$. Here $R = \log\gamma - \frac{r}{2} + \log|\sin\theta|$ and $S = \phi$, so

$$u\cdot\nabla = \left(-\frac{1}{2}+\frac{1}{r}\right)\frac{\partial}{\partial r} - \cot\theta\,\frac{\partial}{\partial\theta}, \quad v\cdot\nabla = \frac{\partial}{\partial\phi}.$$

The nodal set is the x^3-axis. The electron tends to move away from the x^3-axis, towards the sphere $r = 2$, and counterclockwise around the x^3-axis. This process is different from its time reversed process, with wave function $\overline{\psi}_3$, which moves clockwise.

The four functions $\psi_1, \psi_2, \psi_3, \overline{\psi}_3$ are a basis for the eigenspace with eigenvalue $-\frac{1}{8}$, and for any linear combination of them there is a stationary diffusion with expected stochastic energy $-\frac{1}{8}$.

§20. Statistics

Consider N indistinguishable particles in R^n (e.g. $n = 3$). This notion is foreign to macroscopic particle dynamics – how can particles be utterly indistinguishable? But if we think of particles as being a non-relativistic approximation to peaks of a relativistic field, the notion of indistinguishability acquires significance.

The configuration space of N indistinguishable particles in R^n is the set M of all unordered N-tuples $\{x_1, \cdots, x_N\}$ where the x_α are distinct points in R^n. This is a differentiable manifold. It is not simply connected if $N > 1$, because it has as a covering space the manifold $\tilde{M} = R^{nN} \setminus D$, where D is the set of all ordered N-tuples $(x_1, \cdots, x_N)$ in which two or more of the x_α coincide. Each permutation π in the symmetric group $\mathcal{S}_N$ on N objects acts on $\tilde{M}$ and is a covering transformation (the identity on M). Let us assume that $n \geq 3$; then D is of codimension at least 3, and $\tilde{M}$ is simply connected. Then $\tilde{M}$ is the universal covering space of M, and $\mathcal{S}_N$ is the fundamental group of M.

Now consider a Markovian diffusion on M, critical for a basic Lagrangian L. In each region where $\rho \neq 0$, we define $\psi = e^{R+iS}$, with S unique up to a real additive constant (for a fixed gauge), and we impose the local stability requirement of §15 that ψ be smooth across the nodes. Now suppose that we start at a point x in M where $\rho \neq 0$ and continue along a path γ in M, perhaps crossing various nodes, and coming back to x. Since M is not simply connected, we may find that S differs by a real additive constant; i.e. ψ is changed by a multiplicative constant ζ of modulus 1.

The point I want to stress is that such a change in ψ does not affect the diffusion process. The $u^j = \nabla^j R$ and $v^j = \nabla^j S - A^j$ are totally unaffected by an additive constant in S. The diffusion process is well defined on M even though ψ may not be.

The factor ζ depends only on the homotopy class of the path γ, i.e. only on the permutation π in the fundamental group $\mathcal{S}_N$, so that $\zeta = \zeta(\pi)$. Observe that $\zeta(\pi_2\pi_1) = \zeta(\pi_2)\zeta(\pi_1)$; i.e., ζ is a character of the symmetric group $\mathcal{S}_N$. There are only two such: the identity and the sign of the permutation.

In both cases the wave function ψ is well defined on $\tilde{M}$. In the first case (Bose-Einstein statistics) it is a symmetric function on $\tilde{M}$ and in the second case (Fermi-Dirac statistics) it is an antisymmetric function on $\tilde{M}$. Notice that a superposition of a symmetric and an antisymmetric wave function does not define a diffusion on M.

Thus conservative diffusions of indistinguishable particles fall into two sharply different classes according to the statistics obeyed. In one of the appendices to his beautiful book "Subtle is the Lord ...", Pais [50, p. 517] reproduces a telegram from Einstein to the Nobel committee nominating Pauli for his discovery of the exclusion principle, which Einstein describes as a "...fundamental part of modern quantum physics being independent from the other basic axioms of that theory." The exclusion principle appears to follow (once one knows the answer, thanks to Pauli!) from the fundamental assumptions of stochastic mechanics.

I think there is a moral to be drawn here. Probabilistic techniques have not yet been applied in a natural way to the study of Fermi fields, either in imaginary time or real time. My hunch is that no departure from the framework of stochastic mechanics is needed; that one needs to find the right classical configuration space and study ordinary diffusions, Wiener process plus a drift.

§21. Spin

Let M be a Riemannian manifold, thought of as the configuration space of a particle. The main example is $\mathbf{R}^3$ with mass tensor $m\delta_{ij}$, but some conceptual issues will be clearer if we discuss the general case — for one thing, we will be spared the degradation of confusing 2-forms with vectors and the horror of the vector product.

Let n be the dimensionality of M. The orientation of a particle at x in M is described by a frame, by which I mean an orthonormal basis $y = (\vec{y}_1, \cdots, \vec{y}_n)$ of T_xM. The orientation may change while the particle stays at x. There is no implication of an extended structure in this kinematical picture: a point particle can just as easily rotate 30° in a plane as can an extended body. For some extended bodies, e.g. an asymmetrical molecule to the approximation that it can be treated as a rigid body, the orientation can be observed, but there is no way of doing this for a point particle — then the effects of the orientation component of the motion can be observed only indirectly through a coupling to the position component. (It is important to distinguish orientation from the time rate of change of orientation, which is spin.) Following Dankel [9], we will find that for conservative diffusions there is such a coupling, in contrast to deterministic mechanics.

The set of all frames is a differentiable manifold P, the frame bundle. It is a fiber bundle over M with projection $\pi: P \to M$, and the fiber at x is the set P_x of all frames at x. The structural group is the orthogonal group $O(n)$: if y is a frame at x, then any other element of P_x is of the form gy for a unique g in $O(n)$. If M is orientable, then P has two disconnected components, and we can choose one of them and reduce the structural group to $SO(n)$. However, I do not want to assume that M is orientable. Opinions differ, but in my opinion it is a flaw in a local dynamical theory if the global formulation of the theory is only possible with global restrictions on the topology of space. For a basic Lagrangian L on M, let us seek to extend it taking orientation into account, in a way that is intrinsic (i.e. functorial, uniquely determined by L, with no arbitrary choices and with no restrictions on the topology of M).

As we have seen, P_x can be identified with $O(n)$ if we choose an element y in P_x to correspond to the identity in $O(n)$. Thus $O(n)$ is a group and P_x is its principal homogeneous space (a group that has lost its sense of identity). If we can choose a smooth section $y: M \to P$ (so that $\pi \circ y$ is the identity on M), then we may identify P with $M \times O(n)$, but the condition that this be possible (the condition that M be parallelizable) is an even stronger condition than orientability.

Let y be a frame at x. Then there is a natural identification of $T_y P_x$ with the space A_x^2 of 2-forms at x. For if ω_{ij} is in A_x^2, then $\omega^i{}_j$ is the matrix of a skew-symmetric linear transformation, the infinitesimal generator of a rotation, and vice versa. Then $\omega^i{}_j$ is the angular velocity vector in $T_y P_x$. (Notice that P_x is $n(n-1)/2$ dimensional. It is only when $n = 3$ and M is orientable that one can identify $T_y P_x$ with $T_x M$.)

We define an inner product on T_yP_x by setting $<\omega^i{}_j, \eta^k{}_\ell> = \omega^i{}_j \eta^j{}_i$. This inner product is invariant under the action of SO(n) on T_yP_x, and up to a constant factor it is the unique inner product on T_yP_x with this property. Since we have an inner product on T_yP_x for each y, we have a Riemannian metric on P_x. It is also invariant under the action of SO(n) on P_x.

If we choose local coordinates at x in M, then any Y in $T_{xy}P$ has components $(X^i, \omega^i{}_j)$. The splitting into velocity and angular velocity depends on the choice of local coordinates, but we define an intrinsic splitting $T_{xy}P = T_xM \oplus T_yP_x$ by taking normal coordinates at x (expressed invariantly, we use the Riemannian connection on M to define the horizontal subspace in $T_{xy}P$). Then we give T_{xy} the inner product that is the direct sum of the inner products on T_xM and T_yP_x. This defines an intrinsic Riemannian metric on P.

Observe that $\frac{1}{2}\,\omega^i{}_j\omega^j{}_i$ has the dimensions of angular acceleration (T^{-2}). To express the kinetic energy of rotation, we need a coefficient I with the dimensions of moment of inertia (ML^2). This can only be introduced ad hoc, and any natural definition in a nonrelativistic theory of the moment of inertia of a point particle gives it the value 0. Therefore we choose $I > 0$ with the dimensions ML^2 and then study the limit as $I \to 0$.

We will be interested in the case that M is $\mathbf{R}^3$ with the flat metric $m\delta_{ij}$, but the induced metric on P is not flat; this is why we studied the effects of curvature on diffusion. Perhaps the expression $\omega^i{}_j\eta^j{}_i$ looks flat, but the $\omega^i{}_j$ are not the components of the angular velocity field with respect to a coordinate system. For $i < j$, let $\partial_i{}^j$ be the infinitesimal generator of rotation with unit speed in the $\vec{y}_i\vec{y}_j$-plane. Then the $\partial_i{}^j$ are a basis for T_yP_x, and any angular velocity field ω on P_x is expressible uniquely and globally as $\omega^i{}_j\partial_i{}^j$. This is more useful than any system of local coordinates, but notice that the $\partial_i{}^j$ do not in general commute.

Now let L be the basic Lagrangian on M with scalar potential ϕ and covector potential A_i. We want to extend it to a basic Lagrangian $\bar{L}$

on P. We have already extended the kinetic energy part. We extend ϕ to P by setting $\overline{\phi}(x,y) = \phi(x)$. To extend the covector potential, let the generalized velocity Y^i on P have the components p^i and $\omega^i{}_j$, by the splitting described above, and let $\overline{A}_i = (A_i, IH_i{}^j)$ where H_{ij} is the exterior derivative of A_i, so that $\overline{A}_i Y^i = A_i P^i + IH_i{}^j \omega^i{}_j$. Thus

$$\overline{L} = \frac{1}{2} m_{ij} p^i p^j + \frac{1}{2} I\omega^i{}_j \omega^j{}_i - \phi + A_i p^i + IH_i{}^j \omega^i{}_j . \tag{21.1}$$

Notice that in deterministic mechanics the factor I drops out of the orientational component of Newton's equation. The limit $I \to 0$ certainly exists, but it is uninteresting because the positional component is completely decoupled from the orientational component in this limit. This is the meaning of the statement that there is no classical (deterministic) analogue of spin.

The group SO(n) is not simply connected. To see this, take off your belt. Hold the two ends together; this is the standard immersion of the belt in R^3. Now twist one end through 2π, obtaining another immersion. If this twist were homotopic to the identity, the new immersion would be isotopic to the first. But try: you may twist the belt (keeping the ends together) or slip portions of the belt between the two ends, but you will never recover the standard immersion. Now do the same thing with a twist of 4π and notice the difference.

The group SO(n) has a double covering group Spin(n), which for $n \geq 3$ is simply connected. For $n = 3$ it is isomorphic to SU(2). If we identify either component of P_x with SO(n) by choosing a frame y to correspond to the identity, the double covering we obtain is independent of the choice of y. Thus we have a double covering $\widetilde{P} \to P$. We lift the Riemannian metric to $\widetilde{P}$. If M is R^3, we may take P to be $R^3 \times SO(3)$ and $\widetilde{P}$ to be $R^3 \times SU(2)$ (discarding the other connected component).

I want to stress that the configuration space is P, not $\widetilde{P}$, and we will study diffusions on P. I cannot see that diffusions on $\widetilde{P}$ are of any physical significance. But since P is not simply connected, the wave function of a conservative diffusion on P need not be defined on it, just as in the discussion of statistics in the previous section. If y is in $\widetilde{P}$, let $-y$ be the other point in $\widetilde{P}$ with the same projection onto P. Then the wave functions on $\widetilde{P}$ that correspond to diffusions on P split into two classes: *integral spin wave functions* satisfying $\psi(x,y) = \psi(x,-y)$ *and half-integral spin wave functions* satisfying $\psi(x,y) = -\psi(x,-y)$. A superposition of wave functions of the two classes does not correspond to a diffusion on configuration space.

The Schrödinger equation on $\widetilde{P}$ corresponding to the Lagrangian $\overline{L}$ is the *Bopp-Haag equation*

$$i\hbar \frac{\partial\psi}{\partial t} = H_I\psi \tag{21.2}$$

where H_I is the time dependent Hamiltonian operator

$$H_I = \frac{1}{2}\left(\frac{\hbar}{i}\nabla^j - A^j\right)\left(\frac{\hbar}{i}\nabla_j - A_j\right) + \frac{1}{2I}\left(\frac{\hbar}{i}\partial_j{}^k - IH_j{}^k\right)\left(\frac{\hbar}{i}\partial^j{}_k - IH^j{}_k\right) + \phi\,, \tag{21.3}$$

with the convention that $j < k$. If the potentials are suitably smooth, and with appropriate boundary conditions on M if relevant, H_I is a self-adjoint operator on $L^2(\widetilde{P})$ with a time independent domain, and the initial value problem for (21.2) has a unique solution.

Now $\partial^2 = \partial_j{}^k\partial^j{}_k$ is the Laplacian on $\widetilde{P}_x$. Since $\widetilde{P}_x$ is compact, $L^2(\widetilde{P}_x)$ is the direct sum of the eigenspaces $\mathcal{H}_\lambda$ of $-\frac{1}{2}\partial^2$, and each $\mathcal{H}_\lambda$ is invariant under the action of $\mathrm{Spin}(n)$. There is no intrinsic way of splitting $\mathcal{H}_\lambda$ into a direct sum of irreducible subspaces; if it is possible to do so smoothly as x varies over M, then M is said to admit a spin structure. For $n = 3$ the eigenvalues of $-\frac{1}{2}\partial^2$ are $s(s+1)$ for $s = 0, \frac{1}{2}, 1, \frac{3}{2}, \cdots$, the corresponding eigenspace is $(2s+1)^2$ dimensional, and the representation of $SU(2)$ on it is the $(2s+1)$-fold direct sum of the irreducible spin s representation.

Let $\mathcal{K}_\lambda$ be the subspace of $L^2(\widetilde{P})$ such that for a.e. x in $M, y \mapsto \psi(x,y)$ is in $\mathcal{H}_\lambda$. We have seen that for kinematical reasons ψ must be either an integral or a half-integral spin wave function. Now let us show that in the limit $I \to 0$, for dynamical reasons ψ must be in $\mathcal{K}_\lambda$ for some λ (so that for $n = 3$ it has a definite spin $0, \frac{1}{2}, 1, \frac{3}{2}, \cdots$). Let

$$H_o = \frac{1}{2}\left(\frac{\hbar}{i}\nabla^j - A^j\right)\left(\frac{\hbar}{i}\nabla_j - A_j\right) + \hbar H_j{}^k \partial^j{}_k + \phi . \tag{21.4}$$

Then

$$H_I\psi = H_o\psi + \frac{\hbar^2\lambda}{I}\psi + \frac{I}{2} H_j{}^k H^j{}_k\psi , \quad \psi \in \mathcal{K}_\lambda . \tag{21.5}$$

Even though $\hbar^2\lambda/I \to \infty$ as $I \to 0$, it contributes a harmless phase factor to the solution of (21.2), and the last term tends to 0 in the operator norm as $I \to 0$. Let ψ^o be in $L^2(\widetilde{P})$, let ψ_I be the solution of (21.2) with $\psi_I(x,0) = \psi^o(x)$, and let ψ_o be the solution of the *Dankel equation*

$$i\hbar \frac{\partial\psi}{\partial t} = H_o\psi \tag{21.6}$$

with $\psi_o(x,0) = \psi^o$. Also, let $[\psi]$ be the equivalence class of all multiples $e^{i\theta}\psi$. Then if ψ^o is in $\mathcal{K}_\lambda$,

$$\lim_{I\to 0} [\psi_I] = [\psi_o] . \tag{21.7}$$

But if ψ^o has components in several different spaces $\mathcal{K}_\lambda$, then $[\psi_I]$ has no limit as $I \to 0$; in particular, the probability density $|\psi_I|^2$ has no limit.

Now suppose that ψ^0 is in the domain of H_I and in $\mathcal{K}_\lambda$ for some λ. Then, using the technique of §15 or [67], we may construct the probability measure Pr_I for the corresponding Markovian diffusion on P. These measures have no limit as $I \to 0$ because of the increasingly wild behavior of the orientational component: the diffusion coefficient is $1/(2I)$, which tends to ∞. But let $\hat{Pr}_I$ be the probability measure on the positional component, induced by the projection $P \to M$.

TRUE ASSERTION 21.1. *If* ψ^0 *is in the domain of* H_I *and in* $\mathcal{K}_\lambda$, *for some* λ, *then* $\lim_{I\to 0} \hat{Pr}_I$ *exists.*

Convincing argument. By (21.7), the osmotic and current momenta u_I and v_I converge, away from the nodes, to the osmotic and current momenta u_o and v_o computed from ψ_o. (The angular momenta have a coefficient I, so that they converge although the angular velocities diverge; see below). Let $\hat{E}_t$ be the conditional expectation with respect to the σ-algebra of events depending only on the positional component at time t. Then we may use $\hat{E}_t u_o$ and $\hat{E}_t v_o$ to construct a Markovian diffusion on M. But for $I > 0$, the positional component is not Markovian. This is a general feature of taking one component of a Markov process: if we know the past history of it, this may give us more information about the present value of the other component. But when I is very small, knowledge of the orientational component is of no use anyway, because the relaxation time is practically instantaneous, being proportional to I. There is one proviso: the nodes may divide the fibers into a finite number of regions G_i with no communication among them, and knowing the past history of the positional component may give us information about which region the orientation is in. Let p_i be the probability that the orientation is trapped in the region G_i; this is time independent. Then the limiting process as $I \to 0$ is the mixture, with weightings p_i, of the Markovian diffusions on M with osmotic and current momenta u_i and v_i, where these are the conditional expectations of u_o and v_o given that the orientation is in G_i. ∎

There is clearly work to be done to make this into a theorem. When M is R^3 with the metric $m\delta_{ij}$, the Dankel equation is just the Pauli equation of spin s but with multiplicity $(2s+1)$, so this procedure should give a natural probabilistic interpretation of the Pauli equation. Let us look at this case more closely, but first I would like to make a comment about the general case.

Let $I > 0$. Then we know that the stochastic Newton equation holds. The positional component of the force corresponding to the Lagrangian $\bar{L}$ is

(21.8) $$F_i = E_i + H_{ij}p^j + \frac{I}{m}\nabla_i(H_j{}^k\omega^j{}_k)\,.$$

Now we substitute the current velocity v^i for p^i and the current angular velocity $\omega_v{}^j{}_k$ for $\omega^j{}_k$. The first two terms on the right-hand side of (21.8) give the usual Lorentz-type force, but for the last term we will have that

(21.9) $$\lim_{I\to 0} \hat{E}_{x,t}\,\frac{I}{m}\nabla_i(H_j{}^k\omega_v{}^j{}_k)$$

exists, in general is not zero, and is a contribution to the stochastic Newton equation that depends on the process; another indication that there is no deterministic analogue of spin. This contribution to the stochastic acceleration is crucial to the motion of a particle with spin in presence of an inhomogeneous magnetic field (the Stern-Gerlach experiment).

Now let $P = \mathbf{R}^3 \times SO(3)$ and $P = \mathbf{R}^3 \times SU(2)$. It aids visualization to recognize that $SO(3)$ is diffeomorphic to real projective 3-space. To see this, consider the ball of radius π in $\mathbf{R}^3$ with opposite points on the boundary identified (this is projective 3-space), and let each point x in it correspond to the rotation through an angle $|x|$ about the axis x. Consequently, $SU(2)$ is diffeomorphic to S^3; in fact, it is isomorphic as a Lie group to the unit quaternions with the Pauli matrices corresponding to i, j, k.

The Euler angles provide a system of local coordinates for an open dense coordinate neighborhood in $SO(3)$. They are ill adapted to computation, but are helpful for visualization. Let g be a rotation that does not take the x^3-axis into its negative. Then we write

$$g = g_\chi g_\theta g_\phi$$

where g_ϕ is a rotation through ϕ about the x^3-axis, g_θ is a rotation through θ about the axis into which g_ϕ takes the x^1-axis, and g_χ is a rotation through χ about the axis into which g_θ takes the x^3-axis.

All three angles lie in $(-\pi,\pi)$. Then θ is the angle by which g displaces the north pole. Let $\bar{\phi}$, $\bar{\theta}$, and $\bar{\chi}$ be unit vectors on the axes of rotation of g_ϕ, g_θ, and g_χ.

Let us discuss free diffusions on SO(3), for $I > 0$, following Dankel. Let $\mathcal{H}^s$ be the eigenspace $\mathcal{H}_{s(s+1)}$. Then $\mathcal{H}^0$ consists of the constants, and the corresponding diffusion process is the Wiener process on SO(3) with diffusion coefficient $1/(2I)$, so it jiggles about faster and faster as $I \to 0$. Spin 0 diffusions are uninteresting, because the positional and orientational motions decouple as $I \to 0$.

A basis for $\mathcal{H}^{1/2}$ is given by

$$(21.10)\quad \begin{cases} \psi_{\frac12\frac12} = e^{\frac{i}{2}(\phi+\chi)}\, i\cos\frac{\theta}{2}, & \psi_{\frac12-\frac12} = -e^{\frac{i}{2}(\phi-\chi)}\, i\sin\frac{\theta}{2}, \\ \psi_{-\frac12\frac12} = e^{\frac{i}{2}(-\phi+\chi)}\sin\frac{\theta}{2}, & \psi_{-\frac12-\frac12} = e^{\frac{i}{2}(-\phi-\chi)}\cos\frac{\theta}{2}. \end{cases}$$

Notice that because of the occurrence of $\frac{\theta}{2}$, these functions are defined not on SO(3) but on SU(2).

Consider the diffusion corresponding to $\psi_{\frac12\frac12}$, and follow the practice of identifying an angular velocity with a vector in 3 dimensions. Then the current angular velocity $\vec{\omega}_v$ and the osmotic angular velocity $\vec{\omega}_u$ are given by

$$(21.11)\qquad \vec{\omega}_v = \frac{1}{1+\cos\theta}\frac{\hbar}{2I}(\bar{\phi}+\bar{\chi}), \quad \vec{\omega}_u = -\frac{\hbar}{2I}\tan\frac{\theta}{2}\,\bar{\theta}.$$

The corresponding angular momenta $I\vec{\omega}_v$ and $I\vec{\omega}_u$ do not depend on I. Notice that $\vec{\omega}_v$ and $\vec{\omega}_u$ are defined on SO(3) since tan has period π rather 2π; the diffusion is on SO(3). The tendency of the osmotic angular velocity is to restore the north pole to its upright position, and this tendency becomes infinitely strong as it approaches the south pole. There is a node at $\theta = \pi$, where the coordinate system breaks down. This is the plane at infinity when SO(3) is regarded as projective

3-space, and the nodes do not disconnect SO(3) in this example. The tendency of the current osmotic velocity is to produce a rotation through the axis bisecting the x^3-axis and the axis into which it has been carried. On all of this is superimposed the rotational Wiener process with diffusion coefficient $1/(2I)$. Picture this in the limit as $I \to 0$: that is what a spinning electron looks like. We have $EI\vec{\omega}_u = 0$, $EI\vec{\omega}_v = \frac{\hbar}{2}\bar{\phi}$, and $E(I\vec{\omega}_v + I\vec{\omega}_u)^2 = \frac{3}{4}\hbar^2$.

Since the nodes do not disconnect SO(3) for the spin $\frac{1}{2}$ Pauli equation, this might be a good place to begin to make Assertion 21.1 into a theorem and obtain a clear picture of the diffusion on R^3 in the limit as $I \to 0$.

The theory of spin in stochastic mechanics, and more generally of stochastic mechanics on a Riemannian manifold, is due to Dankel [9]. The dynamics was clarified by Dohrn and Guerra's discovery [14] [15] of their stochastic parallel translation. Other contributions to the theory include [10], [21], and [16].

Chapter IV
PHYSICS OR FORMALISM?

Stochastic mechanics has a natural derivation from the variational principle, and its predictions – which agree with the predictions of quantum mechanics – are confirmed by experiment. Had the Schrödinger equation been derived in this way before the invention of matrix mechanics, the history of the conceptual foundations of modern physics would have been different. Yet stochastic mechanics is more vulnerable than quantum mechanics, because it is more ambitious: it attempts to provide a realistic, objective description of physical events in classical terms. Stochastic mechanics is quantum mechanics made difficult. A number of problems that were solved by quantum mechanics, or which at least were declared to be non-problems, are reopened. For example, the electron in the ground state process of the hydrogen atom performs a wildly accelerated random motion, so why do we not detect a resulting electromagnetic disturbance? A cheap solution to this problem is to say that when stochastic quantization is also applied to the electromagnetic field, then the predictions of stochastic mechanics agree with those of quantum mechanics, but a detailed picture of the energy balance is called for.

In this chapter some problems of the physical interpretation of the theory are discussed, and the book concludes with some remarks about stochastic fields.

§22. Measurements

The first thing to say about measurements is that the outcome of any conceivable experiment may be expressed in terms of the positions of macroscopic objects. This is a triviality. It does not concern the physical observable being measured, but how we as human beings receive information through our senses. Furthermore, the outcome of any conceivable experiment may be expressed in terms of the position of macroscopic objects, including recording devices, at a single time. Consequently, if the laws of quantum mechanics apply both to the system being measured and to the measuring apparatus, then the predictions of quantum mechanics will be identical with those of Markovian stochastic mechanics, since the positions of all constituents of the system plus apparatus are determined by the probability density $\rho = |\psi|^2$, where ψ is the wave function of the system plus apparatus.

Nevertheless, the two theories are very different conceptually. In stochastic mechanics an observable is a random variable, whereas in quantum mechanics it is a self-adjoint operator. For position observables, the correspondence is very close. Consider the configuration at time t of a system. In stochastic mechanics this is described by the random variable $\xi(t)$. Notice that this random variable, and even the probability space on which it is defined, depends on the process, i.e. on the wave function. In quantum mechanics the configuration at time t is described by the Heisenberg position operator $X(t)$, which is defined on a fixed Hilbert space not depending on the wave function. The connection between the two descriptions is determined by the fact that, for all suitably smooth wave functions, $E\xi(t) = \langle\psi, X(t)\psi\rangle$. I will sometimes write the left-hand side of this equation as $E_\psi \xi(t)$, to emphasize that the expectation depends on the process.

As soon as we leave position observables, this correspondence breaks down. Consider a random variable $f(\xi(t), \xi(s))$ depending on two different times. In general, $E_\psi f(\xi(t), \xi(s))$ does not depend in a sesquilinear way on ψ, so there is no corresponding operator on Hilbert space. In stochastic mechanics one may discuss many observables that cannot be formulated in quantum mechanics. For example, consider a free particle of mass m whose wave function at time 0 is ψ_0. Then the first time that the particle enters the unit ball is a well-defined random variable, but I do not know of any way of formulating a corresponding concept within conventional quantum mechanics (but see [52]). Now consider the momentum of a particle. In quantum mechanics there is no ambiguity: this is described by the Heisenberg momentum operator $P_j(t)$. In stochastic mechanics there are many different random variables that describe different aspects of the notion of momentum: we have b_j, b_{*j}, v_j, and the random variables π_{+j} and π_{-j} of §§16, 18. All of these have the property that for any wave function ψ, the expectation is equal to $\langle\psi, P_j(t)\psi\rangle$. On the other hand, for the osmotic momentum u_j we have

$$E_\psi u_j = \frac{\hbar}{2}\int \frac{\nabla_j \rho}{\rho}\,\rho = 0\,,$$

so that the corresponding self-adjoint operator is 0. Yet it would be a mistake to identify the osmotic momentum with 0. Its square $u^j u_j$ is an important contribution to (twice) the kinetic energy. Furthermore, the osmotic momentum can be measured as follows. Prepare a particle so that its wave functions is ψ, measure the position, and then compute the value of $\hbar\nabla_j|\psi|$ at that position. If this is repeated many times, one obtains a well-defined probability distribution that is by no means concentrated at 0, although its expectation for any ψ will be 0. This type of wave function-dependent measurement is not usually considered in the quantum theory of measurement, but it is quite feasible.

From the point of view of stochastic mechanics, the identification of observables with self-adjoint operators appears to be epistemologically unsatisfactory.

Stochastic mechanics attempts to give a realistic picture of events in the microphysical world, but it is not in the spirit of a deterministic hidden variables theory. According to the background field hypothesis, quantum fluctuations are the result of an interaction with the background field, but in this picture there is an intrinsic randomness in the interaction. All attempts to produce a deterministic classical theory of charged point particles in interaction with the electromagnetic field have, in my opinion, failed—for example, the Dirac-Röhrlich theory [53] posits a pre-acceleration to avoid runaway solutions, and this violates causuality. My hunch is that this classical interaction leads to stochastic fluctuations.

Stochastic mechanics does not attempt to produce hidden variables that determine the outcome of every measurement. Nevertheless, stochastic mechanics may be artificially presented as such a hidden variables theory. We take as a fixed probability space Ω the probability space of the differences $w^i(t) - w^i(s)$ of the Wiener process. Then for a solution ψ of the Schrödinger equation with potentials, we construct the corresponding process ξ by solving the stochastic differential equations

$$d\xi^i(t) = b^i(\xi(t),t)\,dt + dw^i(t), \quad t \geq 0,$$

with initial probability density $|\psi(x,0)|^2$, where b^i is the forward drift, and similarly for $t \leq 0$, using the method of successive approximations discussed in §11. Then all of these stochastic processes are defined on the same probability space Ω. If one knows the trajectory of the Weiner difference process, then one knows the trajectory of ξ. This is a purely formal construction. It says roughly that if one knew what would have happened in the ground state process for the free system, then this knowledge would determine what actually happens for the interaction process in question. As we will see later, this picture violates a necessary requirement of a physically realistic theory. Nevertheless, it is a mathematical hidden variables interpretation of stochastic mechanics and so it gives the same predictions as does quantum mechanics.

There is a beautiful proof of the impossibility of hidden variables due to Kochen and Specker [43] that deserves to be more widely known among physicists. Let me say at the outset that it does not contradict stochastic mechanics because they identify observables with self-adjoint operators, and this remark also applies to the Jauch-Piron proof [38].

Consider a particle of spin 1. If x, y, z is any orthonormal basis for $\mathbf{R}^3$, then the squares of the spin operators, S_x^2, S_y^2, and S_z^2, have eigenvalues 0 and 1, have sum 2, and commute. For any non-zero vector x in R^3, let us identify x with the proposition "if $S^2_{x/|x|}$ is measured, the result is 0." If x, y, and z are orthogonal, then precisely one of them is true (if there exists a hidden variables theory in the sense of Kochen and Specker). Kochen and Specker show that it is impossible to assign truth values to the directions in R^3 satisfying this requirement. Let us follow the simplified argument of Friedberg, as expounded in Jammer's book [37, p. 324]. Now $x+y$, $x-y$, z are orthogonal, and so are $x+z$, $x-z$, y. Hence $x+y$, $x-y$, $x+z$, $x-z$ are not all false (for if so, both y and z would be true). Each of these four vectors is orthogonal to one of $d_1 = y+z+x$ and $d_2 = y+z-x$, so d_1 and d_2 are not both true. That is, if $\alpha = \arccos \frac{1}{3}$, then two vectors making an angle α are not both true. Now let d_3 lie in the xy-plane and make an angle α with y. If d_3 is true, then z is false (since it is orthogonal to d_3) and y is false (since it makes an angle α with d_3); therefore x is true. That is, if $\beta = \pi/2 - \alpha$, then two vectors making an angle β are equivalent. But in R^3 any two vectors can be connected by a chain of five vectors with angle β between successive pairs, so all vectors are equivalent, which is a contradiction.

Now let us consider how one can measure the proposition x. I will not follow Kochen and Specker here (the "spin Hamiltonian" they consider appears to be a phenomenological Hamiltonian for atoms in a crystal; it is not a consequence of the spin 1 Pauli equation). Instead, consider a Stern-Gerlach experiment: turn on an inhomogeneous magnetic field in the direction of x —then x is true if and only if the particle is not deflected. Consider a spin 1 particle as discussed in the previous section, with moment of inertia $I > 0$ but very small. Then if we know the trajectory ω in the absence of external fields, we know the trajectory in the presence of external fields; in particular, we know for each x whether x is true or false. Consequently, by the Kochen-Specker theorem, it is not the case that the truth assignments given by this hidden variables interpretation of stochastic mechanics have the property that for any three orthogonal directions precisely one is true. It would be interesting to see computer-generated pictures of the trajectories corresponding to a fixed free trajectory ω but inhomogeneous magnetic fields in different directions.

As remarked earlier, in stochastic mechanics one can define random variables for which there is no corresponding self-adjoint operator in quantum mechanics. Is it possible to measure such random variables? There is a theory, due to Davidson [11], in which the diffusion tensor σ^{ij} is equal to $z\hbar m^{ij}$, where the dimensionless parameter z can take any positive value. This theory cannot be derived from a variational principle without artificially modifying the Lagrangian, and for $z > 1$ it entails abandoning the "usual assumptions" in the discussion near the beginning of §14; nevertheless, this theory also has the property that all probability distributions at a single time are given by $|\psi|^2$. But for random variables depending on more than one time, the probability distribution depends on z. This shows that the features of stochastic mechanics involving several times, such as the autocorrelation functions discussed in §16, cannot be measured by any known means. So of what use are they?

Let me digress to discuss a beautiful experiment, the Kappler experiment, in dissipative diffusion; see [42]. In this experiment a small mirror, perhaps 1 mm^2 and several molecular layers thick, is suspended by a quartz fiber in a container. The density of air in the container may be varied. As the air molecules strike the mirror, they cause it to perform a Brownian motion — that of a harmonic oscillator, since the restoring force due to the torque of the fiber is linear for the very small angles involved. The angle is measured by shining light on the mirror and measuring the reflected spot some distance away. At all densities, the probability distribution of the angle is the same — it is determined by thermodynamics. But graphs of the motion look very different at different densities — they are roughly sinusoidal at low densities and not at all periodic at high densities. In other words, the random variables depending on a single time are determined by thermodynamics, but those depending on several times require a detailed stochastic theory.

Now suppose that Kappler had not thought of shining light on the mirror. Suppose that the only known way to measure the angle was to use measuring instruments in thermal equilibrium with the system. Then there would be problems: if we bounce a heavy particle off the mirror, this gives the mirror a substantial additional momentum, and so produces an uncertainty in the angle at a later time; but if we bounce a light (i.e. not heavy) particle off the mirror, it will be subject to diffusion with a large diffusion coefficient, and so it does not measure the angle accurately. It does not take much imagination to conceive of physicists developing a theory that denies the reality of the mirror's trajectory and describes the angle by an abstract mathematical object, with the objects for two different times not being jointly measurable.

The moral of this digression is clear: if we want to measure features of a system predicted by stochastic mechanics that involve several times, we need measuring instruments that are not in "quantum equilibrium" with the system.

Quantum theory attempts to establish hegemony over all of physics: according to it, all physical systems are subject to quantum fluctuations because quantum fluctuations are not physically real, being merely a consequence of a conceptual framework of universal applicability. But perhaps it is not so. Perhaps quantum fluctuations are just as real as thermal fluctuations and arise from certain interactions, and perhaps not every interaction is subject to quantum fluctuations. Stochastic mechanics and the background field hypothesis free us from the universal domination of quantum theory and allow us to examine this possibility.

Just as the light in the Kappler experiment is not subject to the thermal fluctuations of the air and the mirror, perhaps gravity is not subject to quantum fluctuations. Present day technology does not permit us to make a movie of the trajectory of an electron in a hydrogen atom by bouncing gravitational waves off it, but the point is a serious one and deserves consideration in the study of the physics of processes in which both gravity and quantum effects contribute significantly.

I am now speaking of matters in which I have no competence, so let me refer you to a preprint by Lee Smolin [58] in which "three independent arguments are given for the conclusion that the distinction between quantum fluctuations and real statistical fluctuations in the state of a system will not be maintained in a theory that gives a correct description of phenomena in which quantum and gravitational effects are both important." The three arguments concern the mixing of quantum and thermal fluctuations under coordinate transformations, the evolution from pure to mixed states in black hole evaporation, and the impossibility of distinguishing experimentally between pure and mixed states of the gravitational field.

§23. Locality

The results of experiments in microphysics are subject to chance, and yet there are correlations in the results of measurements on widely separated particles that have interacted in the past. To understand this has been the principal focus of discussions of the foundations of physics from the time of the Einstein-Podolsky-Rosen paper [20].

Bell's theorem (see [5] [6] [7] [44] [60]) has transformed the discussion. Two spin $\frac{1}{2}$ particles in the singlet state are emitted from a source and travel to widely separated regions A and B. At each of A and B an experimenter performs a Stern-Gerlach experiment to measure whether the particle has spin up or down in one of three coplanar directions making angles of 120° with each other. Then quantum mechanics, and stochastic mechanics, predict that if the two directions are equal, the spins will always be opposite, but if the two directions are unequal, the spins will be opposite $\frac{1}{4}$ of the time. (The actual experiments are performed with photons and polarization measurements, which is very similar.)

According to a deterministic world view, some hidden variables h determine the outcome of the experiment. But locality demands that the spin at A in any direction must not depend on the choice of direction in which the spin at B is measured; otherwise there is an instantaneous transmission of information between two widely separated places (and in a relativistic theory, locality is a consequence of causality: for some observers, the decision to measure the spin at B in a certain direction would affect a measurement at A before the decision is taken). That is, locality requires that the outcome (up or down) of a spin measurement at A in the direction i be a function $f_A(i,h)$ of i and the hidden variables h alone, and similarly for the outcome $f_B(j,h)$ of a spin measurement at B in the direction j. We have

$$f_A(i,h) \neq f_B(i,h) \tag{23.1}$$

for all h. Let us divide the h into two classes: in class I are all those for which the three values of $f_A(i,h)$ are all the same, in class II are all the others. If h is in class I, then the spins will be opposite, by (23.1), no matter what directions i and j are chosen. There are six choices of unequal directions i and j. If h is in class II, then two of these six choices lead to opposite spins. Therefore, if the directions i and j are chosen at random, for those cases in which they are unequal there will be opposite spins at least $\frac{1}{3}$ of the time—which contradicts the predictions and experiment [3]. In other words, determinism is ruled out by the causality principle!

Are these correlations consistent with the background field hypothesis? Let us first put cutoffs on the background field: put the system in a box of side L and consider only field oscillators for momenta smaller than κ. Then there are only finitely many field oscillators, and the system is deterministic. The state of the field oscillators and the two particles gives the hidden variables h. Let us choose L and κ to be enormous. Then technically the interaction is nonlocal: as one of the particles moves, it excites each field oscillator which in turn reacts immediately on the other particle wherever it may be. However, this effect should be exponentially small, with the exponent directly proportional to κ and the distance between the two particles, so practically speaking the interaction is effectively local. Therefore, one might say, the argument given above should be approximately correct: the outcome of a measurement in region A of the spin in the direction i should be approximately a function of i and h alone. But let us examine this more closely.

This deterministic and approximately local interaction must be extremely unstable. The Stern-Gerlach experiments force a choice of spin up or spin down that breaks the initial symmetry of the singlet state. Suppose the two detectors are in the same direction; then no matter how much care the experimenter takes to replicate the same initial conditions, a series of runs shows a random succession of up-down and down-up outcomes. As long as L and κ are fixed, the system is deterministic, and if we know the state of the system in the past, or even at one time, then we know the outcome. But this is just a qualitative mathematical fact, and the precision ε with which we need to know the state may be enormously fine. Practically speaking the interaction is effectively stochastic.

Nevertheless, the interaction is subject to exact conservation laws, and if we observe spin up at B (with the two detectors in the same direction) then we know that we will find spin down at A. Thus an observation at B will give us useful information about the behavior of the system outside the light cone of B, information that is additional to the information we had from observing the initial state of the whole system (unless that observation is carried out with utter precision, finer than ε). Practically speaking the system is effectively not locally causal in the sense of Bell [6]. Notice that there is no question here of signals or causal effects being transmitted from B to A.

Can one show that the variables h are such that the outcome of a measurement in region A of the spin in direction i is approximately a function of i and h alone? To do so, one would need to establish quantitative estimates on the approximate locality of the cutoff interaction and quantitative estimates on the precision with which one needs to know h to determine the outcome, and show that the former effect dominates the latter — but there is no reason to expect this to be true. For fixed L and κ, there is no reason to expect Bell's inequality — that with different directions i and j there will be opposite spins at least $\frac{1}{3}$ of the time — and there is no reason to expect this to be true in the limit as L and κ tend to infinity.

This is the picture we may expect if the background field hypothesis is correct: the equations of motion of the total field, background field plus particles, come from a classical local Lagrangian; with cutoffs the systems is deterministic but increasingly unstable as the cutoffs increase; the limiting theory is stochastic; a knowledge of the field in the past does not determine the future behavior of the field; and observation in a region gives information about the behavior of the field outside the light cone of that region, information that is additional to the information obtained from a knowledge of the past (the field is not locally causal in the sense of Bell [6]); but the locality principle holds — if we couple the field to a current in a region, only the behavior of the field in the future light cone of that region will be affected. This picture accords well with what we know about physics.

Let me say a bit more about local causality in the sense of Bell, since the reference [6] is not easily accessible and the point is a crucial one. Let ϕ be a stochastic field; i.e. a stochastic process indexed by smooth test functions f on space-time. For any open set A, let $\mathcal{O}(A)$ be the σ-algebra generated by all $\phi(f)$ where the support of f is contained in A. Let A and B be space-like separated, let Λ be the interior of the intersections of the backward cones of A and B, and let N be contained in the backward cone of A; see Figure 2. Then we may say that ϕ is *locally causal in the sense of Bell* in case

$$\Pr\{\mathcal{O}(A)\,|\,\mathcal{O}(\Lambda \cup N \cup B)\} = \Pr\{\mathcal{O}(A)\,|\,\mathcal{O}(\Lambda \cup N)\}\,.$$

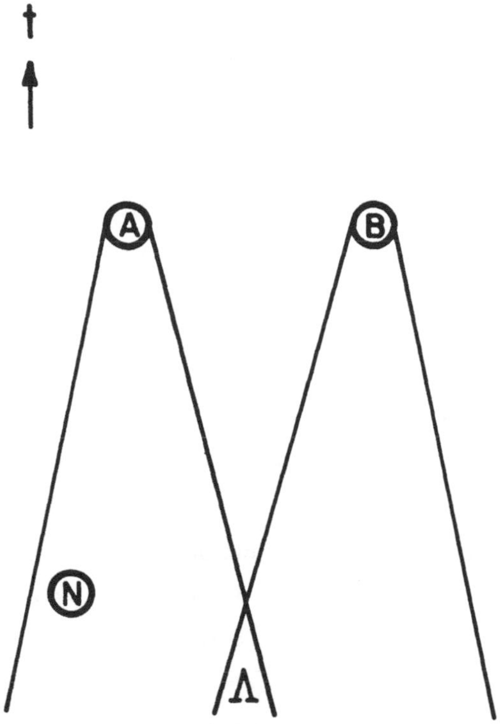

Fig. 2. Local causality in the sense of Bell

In case ϕ is the only relevant field, I think this is a correct exposition of the idea expressed in [6]. The reason Bell gives for imposing the condition is that "... in the particular case that Λ contains already a *complete* specification... in the overlap of the two light cones, supplementary information from region [B] could reasonably be expected to be redundant." Then he proves the beautiful result that quantum mechanics cannot be imbedded in a theory that is locally causal in his sense.

Now suppose that ϕ_κ is a sequence of stochastic fields each of which is locally causal in the sense of Bell, and that $\phi_\kappa \to \phi$ in distribution as $\kappa \to \infty$. For example, each ϕ_κ may be a deterministic field, with randomness entering only in the choice of initial conditions in the remote past. May we conclude that ϕ is also locally causal in the sense of Bell? Not without a proof, and it seems unlikely. The problem is very similar to showing that a limit of Markov fields is a Markov field – for a long time I assumed this to be true, until my mistake was pointed out to me (Lemma 1 of [65] is wrong; see [66]). The point is that in the passage to the limit $\kappa \to \infty$ some information that is present in each $\mathcal{O}_\kappa(\Lambda \cup N)$ may be lost, but some of this information may be retained in $\mathcal{O}(B)$. For each κ, this information may be stored in $\mathcal{O}_\kappa(\Lambda \cup N)$ in terms of extremely precise specification of quantities that are subject to wild fluctuations as κ increases (the hidden variables of a deterministic theory), but it may be stored in $\mathcal{O}(B)$ in terms of a stable configuration (the outcome of a spin measurement). Given $\mathcal{O}(\Lambda \cup N)$, the information in $\mathcal{O}(B)$ may be non-redundant without in any way propagating a causal influence to A.

Bell's theorem and experiment rule out determinism – there is an intrinsic randomness in nature that is not due to our ignorance of initial conditions – but they in no way preclude a treatment of quantum fluctuations as being physically real.

I am not sufficiently versed in philosophy to give a precise definition of "physically real." But the notion has consequences. If something is physically real, then it cannot be affected instantaneously by a widely separated perturbation. This is the locality principle, and it poses a severe challenge to stochastic mechanics. This is because the diffusion occurs on configuration space, and if we have several particles, possibly widely separated, the component of the drift for any particular particle will in general be a function of the positions of all the particles.

Consider a configuration space M that is a Cartesian product, $M = M_1 \times M_2$, and consider an initial wave function ψ_0. Let the potentials split into scalars ϕ_1, ϕ_2 and covector fields A_1, A_2 defined on M_1 and M_2 separately. Let Pr be the probability measure on path space $\Omega = \Pi M$ for the corresponding Markovian diffusion, and let $\Pr_1$ be the probability measure on $\Omega_1 = \Pi M_1$ induced by the projection $M \to M_1$. Then locality requires that $\Pr_1$ not depend on the choice of ϕ_2, A_2. (If M_1 and M_2 are R^3, the configuration space of a particle, then there is no necessary connection between the choices of origin in M_1 and M_2; the two particles may be arbitrarily far apart.) Does Markovian stochastic mechanics satisfy this locality requirement?

To be specific, consider two particles of unit mass in $\mathbf{R}^1$. The first particle is free, and the second particle is a harmonic oscillator of circular frequency ω. Let $x = \begin{pmatrix} x^1 \\ x^2 \end{pmatrix}$, and let the initial wave function be

$$\psi(0) = \frac{1}{\sqrt{2\pi}} e^{-\frac{1}{4}\sigma^{-1}(0)x\cdot x}$$

where $\sigma(0) = \begin{pmatrix} 2 & 1 \\ 1 & 1 \end{pmatrix}$, $\sigma^{-1}(0) = \begin{pmatrix} 1 & -1 \\ -1 & 2 \end{pmatrix}$. The particles are correlated but dynamically uncoupled, and they may be arbitrarily widely separated (the origins on the x^1-axis and x^2-axis are unrelated). Let $\xi = \begin{pmatrix} \xi^1 \\ \xi^2 \end{pmatrix}$ be the corresponding Markov process; the question is whether the ξ^1 process depends on the choice of ω.

We use the notation of §16. The Heisenberg operators are

$$X^1(t) = X^1(0) + tP^1(0), \qquad P^1(t) = P^1(0),$$

$$X^2(t) = \cos\omega t X^2(0) + \frac{\sin\omega t}{\omega} P^2(0), \quad P^2(t) = -\omega\sin\omega t X^2(0) + \cos\omega t P^2(0).$$

Let $\sigma(t) = \langle X(t)\,X(t)\rangle$, $\delta(t) = \langle X(t)\circ P(t)\rangle$, $\tau(t) = \sigma^{-1}(t)\delta(t)$, $\alpha(t) = -\frac{1}{2}\sigma^{-1}(t) + \tau(t)$. The autocorrelation function is

$$E\xi(t)\,\xi(s) = \mathcal{T} e^{\int_s^t \alpha(r)dr}\sigma(s), \quad s \le t.$$

We want to know whether the upper left entry of this depends on ω. Notice that $\tau(0) = 0$, $\delta(0) = 0$, and $\langle P(0)\,P(0)\rangle = \frac{1}{4}\sigma^{-1}(0) = \frac{1}{4}\begin{pmatrix} 1 & -1 \\ -1 & 2 \end{pmatrix}$. Then

$$\sigma(s) = \begin{pmatrix} 2 + \frac{s^2}{4} & \cos\omega s - \frac{s}{4}\frac{\sin\omega s}{\omega} \\ \cos\omega s - \frac{s}{4}\frac{\sin\omega s}{\omega} & \cos^2\omega s + \frac{1}{2}\frac{\sin^2\omega s}{\omega^2} \end{pmatrix}.$$

Since the $\alpha(r)$ do not commute, I see no way to compute the product integral explicitly. John Lafferty and I have computed $E\xi^1(t)\xi^1(0)$ to fourth order in t, and each of us found that due to remarkable cancellations it is independent of ω to that order.

Let $\phi(t) = E\xi(t)\xi(0)$. If $\omega = 0$ and we rotate the axes so that $\sigma^{-1}(0)$ becomes diagonal, the two new coordinates are uncorrelated and dynamically uncoupled, so that the autocorrelation function is the diagonal matrix consisting of the autocorrelation functions for the one-slit process. Therefore in the original coordinates, for $\omega = 0$ we have

$$\phi_{11}(t) = C_\lambda \phi_\lambda + C_{\lambda^{-1}} \phi_{\lambda^{-1}} \tag{23.2}$$

where for any κ

$$\phi_\kappa = \sqrt{4\kappa^2 + t^2}\, e^{-\arctan \frac{t}{2\kappa}},$$

$\lambda = \frac{1}{2}(3+\sqrt{5})$ and $\lambda^{-1} = \frac{1}{2}(3-\sqrt{5})$ are the eigenvalues of $\sigma^{-1}(0)$, and C_λ and $C_{\lambda^{-1}}$ are constants that may be computed from the initial conditions. If locality holds, then we must have (23.2) for all ω. Let us suppose this.

Let $\beta(t) = (\det \sigma(t))\alpha(t)$. Then

$$\det \sigma \begin{pmatrix} \phi_{11}' \\ \phi_{12}' \end{pmatrix} = \begin{pmatrix} \beta_{11} & \beta_{12} \\ \beta_{12} & \beta_{22} \end{pmatrix} \begin{pmatrix} \phi_{11} \\ \phi_{12} \end{pmatrix},$$

$$\phi_{12} = \frac{\det \sigma (C_\lambda \phi_\lambda' + C_{\lambda^{-1}} \phi_{\lambda^{-1}}') - \beta_{11}(C_\lambda \phi_\lambda + C_{\lambda^{-1}} \phi_{\lambda^{-1}})}{\beta_{12}},$$

$$\det \sigma \left(\frac{\det \sigma (C_\lambda \phi_\lambda' + C_{\lambda^{-1}} \phi_{\lambda^{-1}}') - \beta_{11}(C_\lambda \phi_\lambda + C_{\lambda^{-1}} \phi_{\lambda^{-1}})}{\beta_{12}} \right)' \tag{23.3}$$

$$= \beta_{12}(C_\lambda \phi_\lambda + C_{\lambda^{-1}} \phi_{\lambda^{-1}}) + \frac{\beta_{22}}{\beta_{12}} \det \sigma (C_\lambda \phi_\lambda' + C_{\lambda^{-1}} \phi_{\lambda^{-1}}') - \frac{\beta_{22}}{\beta_{12}} \beta_{11}(C_\lambda \phi_\lambda + C_{\lambda^{-1}} \phi_{\lambda^{-1}})$$

Now perform the differentiations, using the facts that

$$\phi'_\kappa = \frac{t-2\kappa}{4\kappa^2+t^2}\,\phi_\kappa ,$$

$$\phi''_\kappa = \frac{8\kappa^2}{(4\kappa^2+t^2)^2}\,\phi_\kappa ,$$

and multiply through by β_{12}^2. Now $\det\sigma$ and β are entire functions of t —computation shows that

$$\beta_{11}(t) = \left(\frac{t}{4}-\frac{1}{2}\right)\cos^2\omega t + \left(\frac{t}{16\omega^2}-\frac{1}{4\omega^2}\right)\sin^{-2}\omega t + \frac{1}{4\omega}\cos\omega t\sin\omega t ,$$

$$\beta_{12}(t) = \left(\frac{1}{2}-\frac{t}{4}\right)\cos\omega t - \left(\frac{t}{8\omega}+\frac{1}{2\omega}\right)\sin\omega t ,$$

$$\beta_{22}(t) = -1-\frac{t^2}{8}+\left[-\left(1+\frac{t^2}{4}\right)\omega+\left(1+\frac{t^2}{16}\right)\frac{1}{\omega}\right]\cos\omega t\sin\omega t - \frac{t}{4}\sin^2\omega t+\frac{t}{4}\cos^2\omega t .$$

The term coming from ϕ''_λ in (23.3), namely

$$(\det\sigma)^2 C_\lambda \frac{8\lambda^2}{(4\lambda^2+t^2)^2}\,\phi_\lambda ,$$

has a singularity at $t = 2i\lambda$ that is not cancelled by anything else in the equation, so we must have $\det\sigma(2i\lambda) = 0$, but this is simply not true for all ω. Thus locality fails (although the argument does not tell us to what order in t we must compute $E\xi^1(t)\xi^1(0)$ to find a dependence on ω). That is, the ξ^1 process does depend on ω —the first particle knows what potential has been switched on at the position of the second particle, even though it may be arbitrarily far away.

I have loved and nurtured Markovian stochastic mechanics for 17 years, and it is painful to abandon it. But its whole point was to construct a physically realistic picture of microprocesses, and a theory that violates locality is untenable.

As remarked earlier, there is no reason why the diffusion of a system produced by interaction with the background field should be Markovian. Markov processes are simple to construct mathematically, but they have the great flaw that a component of a Markov process (such as ξ^1 in the example above) need not be a Markov process. Let us see whether locality is restored if we consider non-Markovian diffusions.

Let us say that a diffusion process on M is *associated* with a solution ψ of the Schrödinger equation in case the probability density at all times is $|\psi|^2$ and the drifts are given by (14.35). The diffusion processes associated with ψ form an equivalence class that contains a unique Markovian diffusion (with Neumann boundary conditions, if relevant). This Markovian approximation determines the equivalence class.

Again let $M = M_1 \times M_2$ and let the potentials split (as above), and let $\xi = (\xi^1, \xi^2)$ be a diffusion associated with a solution ψ of the Schrödinger equation on M. Then the solution of the Schrödinger equation is given by

$$\psi(t) = [U_1(0,t) \otimes U_2(0,t)]\psi(0) ,$$

where U_1 is the unitary propagator on $\mathcal{H}_1 = L^2(M_1, d_{M_1}x^1)$ for the potentials ϕ_1, A_1, and U_2 is defined similarly on $\mathcal{H}_2 = L^2(M_2, d_{M_2}x^2)$. Let A_1 be a self-adjoint operator on $\mathcal{H}_1$. Then the Heisenberg operator

$$A_1(t) = [U_1(t,0) \otimes U_2(t,0)]\,[A_1 \otimes 1]\,[U_1(0,t) \otimes U_2(0,t)] = U_1(t,0)A_1U_1(0,t) \otimes 1$$

does not depend on ϕ_2, A_2. (Since self-adjoint operators are identified with observables in quantum mechanics, quantum mechanics satisfies the locality principle in this way.) In particular, if A_1 is multiplication by $f(x^1)$, then

$$\langle \psi(0), U_1(t,0) f(x^1) U_1(0,t)\psi(0)\rangle = \langle \psi(t), f(x^1)\psi(t)\rangle = E\, f(\xi^1(t)) =$$

$$\iint f(x^1)\rho(x^1,x^2,t)\,d_{M_1}x^1 d_{M_2}x^2 = \int f(x^1)\rho_1(x^1,t)\,d_{M_1}x^1$$

where $\rho_1(x^1,t) = \int \rho(x^1,x^2,t)\,d_{M_2}x^2$ is the probability density at time t of ξ^1. Since f is arbitrary, ρ_1 does not depend on ϕ_2, A_2.

Now $u^j(x^1,x^2,t) = \frac{1}{2}\nabla^j \log\rho$, so that $u^j = (u_1{}^j, u_2{}^j)$ where $u_1{}^j = \frac{1}{2}\nabla_1{}^j \log\rho$ and similarly for $u_2{}^j$. Let

$$\bar{u}_1{}^j(x^1,t) = \frac{1}{\rho_1}\int u_1{}^j(x^1,x^2,t)\,d_{M_2}x^2$$

$$= \frac{1}{\rho_1}\int \frac{1}{2}(\nabla_1{}^j \log\rho)\,\rho(x^1,x^2,t)\,d_{M_2}x^2 = \frac{1}{\rho_1}\frac{1}{2}\nabla_1{}^j\rho_1 = \frac{1}{2}\nabla_1{}^j \log\rho_1 .$$

Then $\bar{u}_1{}^j$ also does not depend on ϕ_2, A_2.

Now let $A_1{}^j = f(x^1)\left(\frac{1}{i}\nabla_1{}^j\right)f(x^1)$ where $f(x^1)$ is a real multiplication operator. Then

$$<\psi(0), U_1(t,0)f(x^1)\left(\frac{1}{i}\nabla_1{}^j\right)f(x^1)U_1(0,t)\psi(0)> = <\psi(t), f(x^1)\left(\frac{1}{i}\nabla_1{}^j\right)f(x^1)\psi(t)>$$

$$= \iint e^{R-iS} f(x^1)\left(\frac{1}{i}\nabla_1{}^j\right)f(x^1)e^{R+iS}\,d_{M_1}x^1 d_{M_2}x^2$$

is a real number not depending on ϕ_2, A_2. The only part that is not purely imaginary is

$$\iint e^{R-iS} f(x^1)\nabla_1{}^j S\, f(x^1)e^{R+iS}\,d_{M_1}x^1 d_{M_2}x^2$$

$$= \iint f^2(x^1)v_1{}^j(x^1,x^2,t)\,\rho(x^1,x^2,t)\,d_{M_1}x^1 d_{M_2}x^2$$

$$= \int f^2(x^1)\bar{v}_1{}^j(x^1,t)\,\rho_1(x^1,t)\,d_{M_1}x^1 = E f^2(\xi^1(t))\bar{v}_1{}^j(\xi^1(t),t)$$

where $\overline{v}_1{}^j(x^1,t) = \frac{1}{\rho_1} \int v_1{}^j(x^1,x^2,t)\rho(x^1,x^2,t)\,d_{M_2}x^2 = \frac{1}{\rho_1} \int \nabla_1{}^j S\, e^{2R} d_{M_2}x^2$. Since f is arbitrary, $\overline{v}_1{}^j$ does not depend on ϕ_2, A_2. That is, the probability density and the drifts of ξ^1 do not depend on ϕ_2, A_2. We have proved the following *locality theorem*:

THEOREM 23.1. *Let* $M = M_1 \times M_2$ *and let the potentials split. Then the class of* M_1 *components of diffusions associated with a solution of the Schrödinger equation on* M *does not depend on* ϕ_2, A_2.

Now the problem is to pick the correct representative of the class of diffusions associated with a solution of the Schrödinger equation on M. As the example given earlier in this section shows, the Markovian approximation is not the correct representative. (Even apart from its failure to satisfy the locality principle, it would be odd to assume that a particle performs a Markovian diffusion unless at some time in the past it happened to become correlated with another particle, in which case it performs a component of a Markovian diffusion.) As these notes give ample evidence, this course is being given at a time when my thinking about stochastic mechanics is shifting from a Markovian to a non-Markovian framework. At present I can only speculate about what the correct diffusion might be. One possibility is that within the equivalence class there is a unique diffusion that is an absolute minimum of the Yasue action

$$E \int_{t_0}^{t_1} \left[\frac{1}{2}(v^j v_j + u^j u_j) - \phi + A_i v^i\right] dt \,.$$

If so, the locality principle would be satisfied by it, by virtue of the locality theorem.

§24. Fields

The first to investigate fields from the viewpoint of stochastic mechanics were Guerra and Ruggiero [29]. They treated the cutoff free scalar field as an assembly of independent harmonic oscillators, each of which performs the Markovian diffusion associated with the ground state wave function, and then removed the cutoffs. They found that the limiting random field is the free Euclidean Markov field. This gives a new interpretation of the free Euclidean Markov field, in real time — see the remarks in §19.

The problem has the Poincaré group as symmetry group but the solution does not; the symmetry is broken. The free Euclidean Markov field in any other Lorentz frame is an equally valid solution. One ought to be able to see this working in a fixed frame: although the cutoff problem has a unique ground state wave function, this does not ensure uniqueness in the limit. There is no symmetry breaking in the ground state for the *quantized* free scalar field. From the free Euclidean Markov field one should be able to construct the Wightman field operators (possessing Poincaré symmetry) by analogy with the construction of the Heisenberg position operators in the ground state process of a system of finitely many degrees of freedom (see §19).

Other integral spin Bose fields have been studied from the viewpoint of stochastic mechanics (see [12] [13] [27]), but the half integral spin Fermi fields have not yet received an adequate treatment (in real or imaginary time). The work of Dankel should provide a clue. In a relativistic theory there is a natural definition of moment of inertia: the mass times the square of the Compton wavelength, $\hbar^2/mc^2$, or a dimensionless constant λ times this. It is possible that one should not take the limit $\lambda \to 0$ but allow for transitions to different spins at relativistic energies.

I have talked for a month about probability theory and quantum theory without once mentioning the Feynman-Kac formula. This formula is at the heart of the imaginary time approach, but it does not appear to be central to the real time approach. In the imaginary time approach, one uses the exponential of the action to obtain a new weighting of the trajectories. In constructive Euclidean quantum field theory this leads to severe problems with divergent quantities, which have been controlled by heroic efforts for many superrenormalizable interactions. In the real time approach, only the variation of the action is relevant and it is not exponentiated, but to reconstruct the Wightman fields one must consider a sufficiently large class of excited states and not just the vacuum. This suggests that stochastic mechanics may provide a useful mathematical tool in constructive quantum field theory, whatever one may think of it as physics.

I have made some preliminary calculations using this approach to nonrelativistic quantum electrodynamics (finitely many Schrödinger particles coupled to the quantized Maxwell field). This theory is much simpler than relativistic quantum electrodynamics because there is no particle creation or vacuum polarization, but the action is much too singular for one to be able to treat its exponential in any simple way. The theory has neither Poincare nor Galilei symmetry, and I do not know of a solution even at the level of formal renormalized perturbation theory. The preliminary calculations are promising, but this material is not at a stage where I can report on it.

So far I have been discussing the application of the methods of stochastic mechanics to fields, with random fluctuations viewed as arising from some background interaction. But the background field hypothesis is that these quantum fluctuations are themselves the result of a classical field interaction. The long range goal of the theory is the Einstein program of describing physical phenomena, including quantum effects, in terms of a classical field theory. Whether this is possible, no one knows. The successes of stochastic mechanics show that, contrary to a widespread belief among physicists, it is not obviously impossible.

I simply do not know whether the things I have been talking about are physics or formalism. But two things I cling to: a belief in external reality, and radical skepticism about all theories of it.

A LIST OF OPEN PROBLEMS

1) To find a classical Langrangian, of system + background field oscillators + interaction, that with reasonable initial probability measures and in the limit as the cutoffs on the background field are removed, produces a conservative diffusion in the system. (Or to show that this is impossible, with a similar proviso for the following problems.) See [69].

2) Same as 1, with Langrangian expressing an electromagnetic interaction.

3) Same as 2, with an explanation of the value of $\hbar$ in terms of e and c.

4) To investigate stochastic mechanics when the diffusion tensor σ^{ij} is degenerate, and the quantization of systems with nonholonomic contraints.

5) To further investigate stochastic mechanics when the forces do not come from a potential, and the quantization of dissipative systems. See [63], [64], [54], [55], [56], [70].

6) To find the probabilistic meaning of the stochastic acceleration.

7) To find the probabilistic meaning, if any, of superpositions and the relative phase factor.

8) To prove the continuity of paths without the Markovian assumption.

9) To find a probabilistic approach (in contrast to Carlen's partial differential equations approach [67]) to the existence and uniqueness of Markovian diffusions of finite energy, and to discover whether with this hypothesis the nodes are ever reached.

10) To investigate scattering theory from the viewpoint of stochastic mechanics.

11) To understand the limit $I \to 0$ in Dankel's theory of spin, and the probabilistic meaning of the Pauli equations.

12) To find a physical conservative diffusion, perhaps in the study of superfluid helium or solid state physics, with an "effective Planck's constant" that is large enough so that one may observe the fluctuations predicted by stochastic mechanics without disturbing the process.

13) To formulate stochastic mechanics within the context of general relativity.

14) To investigate the existence and uniqueness, within the equivalence class of diffusion processes associated with a solution of the Schrödinger equation, of one with an absolute minimum of the Yasue action. See [71].

15) To find a stochastic field theory of half integral spin Fermi fields using ordinary diffusion theory.

16) To study nonrelativistic quantum electrodynamics by the methods of stochastic mechanics.

BIBLIOGRAPHY

This bibliography is limited to works cited in the text, in the sections indicated, with the exception of two groups of papers ([39] [40] [41] and [54] [55] [56] [70]) that treat important topics not covered in the course.

[1] Sergio Albeverio and Raphael Høegh-Krohn, A remark on the connection between stochastic mechanics and the heat equation, J. of Mathematical Physics *15*, 1974, 1745-1747. (§§11, 19)

[2] Sergio Albeverio, Raphael Høegh-Krohn, and Ludwig Streit, Energy forms, Hamiltonians and distorted Brownian paths, J. of Mathematical Physics *18*, 1977, 907-917. (§11)

[3] Alain Aspect, Jean Dalibard, and Gérard Roger, Experimental test of Bell's inequalities using time-varying analyzers, Phys. Rev. Lett. *49*, 1982, 1804-1807. (§23)

[4] J. Azéma and M. Yor, eds., Séminaire de Probabilités XVI, 1980/81, Supplément: Géometrie Différentielle Stochastique, Lecture Notes in Math. *921*, 1982, Springer. (§5)

[5] J. S. Bell, On the Einstein Podolsky Rosen paradox, Physics *1*, 1964, 195-200. (§23)

[6] ———, The theory of local beables, CERN preprint TH-2053, 1975; reproduced in Epistemological Letters (Association Ferd. Gonseth, CP1081, CH-205, Bienne) *9*, 1976, 11. (§23)

[7] ———, Bertlmann's socks and the nature of reality, Journal de Physique *42*, 1981, Suppl., Colloque C2, 41-61. (§23)

[8] G. D. Birkhoff, Relativity and Modern Physics, Harvard University Press, 1923. (§5)

[9] Thaddeus George Dankel, Jr., Mechanics on manifolds and the incorporation of spin into Nelson's stochastic mechanics, Archive for Rational Mechanics and Analysis *37*, 1970, 192-222. (§21)

[10] Thaddeus George Dankel, Jr., Higher spin states in the stochastic mechanics of the Bopp-Haag spin model, J. of Mathematical Physics *18*, 1977, 253-255. (§21)

[11] Mark Davidson, A generalization of the Fényes-Nelson stochastic model of quantum mechanics, Lett. in Math. Phys. *3*, 1979, 271-277. (§22)

[12] ————, Stochastic quantization of the linearized gravitational field, J. of Mathematical Physics, *23*, 1982, 132-137. (§24)

[13] Silvio De Siena, Francesco Guerra, and Patrizia Ruggiero, Stochastic quantization of the vector meson field, Phys. Rev. D, *27*, 1983, 2912-2915. (§24)

[14] D. Dohrn and F. Guerra, Nelson's stochastic mechanics on Riemannian manifolds, Lettere al Nuovo Cimento *22*, 1978, 121-127. (§10)

[15] ————Geodesic correction to stochastic parallel displacement of tensors, in Stochastic Behavior in Classical and Quantum Hamiltonian Systems, ed. G. Casati and J. Ford, Springer Lecture Notes in Physics *93*, 1979, 241-249. (§10)

[16] Daniels Dohrn, Francesco Guerra, and Patrizia Ruggiero, Spinning particles and relativistic particles in the framework of Nelson's stochastic mechanics, in Feynman Path Integrals, ed. S. Albeverio, Lecture Notes in Physics *106*, 1979, Springer. (§21)

[17] John D. Dollard and Charles N. Friedman, Product Integration with Applications to Differential Equations, Encyclopedia of Mathematics and its Applications *10*, 1979, Addison-Wesley, Reading, Mass. (§16)

[18] J. L. Doob, Stochastic Processes, 1953, Wiley, New York. (§4)

[19] E. B. Dynkin, Diffusions of Tensors, Sov. Math. Doklady *9*, 1968, 532-535. (§10)

[20] A. Einstein, B. Podolsky, and N. Rosen, Can quantum-mechanical description of reality be considered complete? Phys. Rev. *47*, 1935, 777-780. (§23)

[21] William G. Faris, Spin correlation in stochastic mechanics, Foundations of Physics *12*, 1982, 1-26. (§21)

[22] Imre Fényes, Eine wahrscheinlichkeitstheoretische Begründung und Interpretation der Quantenmechanik, Zeitschrift für Physik *132*, 1952, 81-106. (§14)

[23] R. P. Feynman, Space-time approach to non-relativistic quantum mechanics, Review of Modern Physics *20*, 1948, 367-387. (§8)

[24] Richard P. Feynman and Albert R. Hibbs, Quantum Mechanics and Path Integrals, 1965, McGraw-Hill, New York. (§17)

[25] D. L. Fisk, Quasi-martingales and stochastic integrals, Tech. Reports *1*, Dept. Math. Michigan State Univ., 1963. (§8)

[26] R. Gangolli, On the construction of certain diffusions on a differentiable manifold, Zeitschrift für Wahrscheinlichkeitstheorie 2, 1964, 406-419. (§11)

[27] Francesco Guerra and Maria I. Loffredo, Stochastic equations for the Maxwell field, Lettere al Nuovo Cimento *27*, 1980, 41-45. (§24)

[28] Francesco Guerra and Laura M. Morato, Quantization of dynamical systems and stochastic control theory, Phys. Rev. D, 1983, 1774-1786. (§§13, 14)

[29] Francesco Guerra and Patrizia Ruggiero, A new interpretation of the Euclidean-Markov field in the framework of physical Minkowski space-time, Phys. Rev. Letters *31*, 1973, 1022-1025. (§24)

[30] W. Heisenberg, The Physical Principles of the Quantum Theory, 1930, University of Chicago Press, Chicago; Dover, New York. (§14)

[31] N. Ikeda and S. Manabe, Integral of differential forms along the path of diffusion processes, Publ. RIMS, Kyoto Univ. *15*, 1979, 827-852. (§8)

[32] Nobuyuki Ikeda and Shinzo Wanatabe, Stochastic Differential Equations and Diffusion Process, North Holland, Amsterdam, 1981. (§5)

[33] Kiyosi Itô, On stochastic differential equations, Memoirs of the Am. Math. Soc. *4*, 1951. (§8)

[34] ————, The Brownian motion and tensor fields on a Riemannian manifold, Proc. Int. Congress Math. (Stockholm), 1962, 536-539. (§10)

[35] ————, Stochastic parallel displacement, in Probabilistic Methods in Differential Equations, ed. M. A. Pinsky, Lecture Notes in Math. *451*, 1975, 1-7, Springer. (§10)

[36] ————, Stochastic differentials, Applied Math. and Optimization *1*, 1975, 374-381. (§5)

[37] Max Jammer, The Philosophy of Quantum Mechanics – The Interpretations of Quantum Mechanics in Historical Perspective, 1974, Wiley, New York. (§22)

[38] J. M. Jauch and C. Piron, Can hidden variables be excluded in quantum mechanics?, Helvetica Physica Acta *36*, 1963, 827-837. (§22)

[39] G. Jona-Lasinio, F. Martinelli, and E. Scoppola, The semi-classical limit of quantum mechanics: a qualitative theory via stochastic mechanics, Physics Reports *77*, 1981, 313-327.

[40] ————, New approach to the semiclassical limit of quantum mechanics I – Multiple tunnelings in one dimension, Comm. Math. Phys. *80*, 1981, 223-254.

[41] ————, Decaying quantum-mechanical states: An informal discussion within stochastic mechanics, Lettere al Nuovo Cimento *34*, 1982, 13-17.

[42] Eugen Kappler, Versuche zur Messung der Avogadro-Loschmidtschen Zahl aus der Brownschen Bewegung einer Drehwaage, Annalen der Phy. *11*, 1931, 233-256. (§22)

[43] Simon Kochen and E. P. Specker, The problem of hidden variables in quantum mechanics, J. Math. Mech. *17*, 1967, 59-87. (§22)

[44] N. D. Mermin, Bringing home the atomic world: quantum mysteries for anybody, Am. J. Phys. *49*, 1981, 940-943. (§23)

[45] P. A. Meyer, A differential geometric formalism for the Itô calculus, in Stochastic Integrals, Proc. LMS Durham Symposium 1980, ed. D. Williams, Springer Lecture Notes in Math. *851*, 1981, 256-270. (§§5, 8, 9)

[46] Edward Nelson, Regular probability measures on function space, Ann. Math. *69*, 1959, 630-643. (§3)

[47] ————, Dynamical Theories of Brownian Motion, 1967, Princeton University Press. (§§8, 10, 14)

[48] ————, Tensor Analysis, 1967, Princeton University Press. (§5)

[49] ————, Critical diffusions, to appear in Séminaire de Probabilités, vol. XIX, ed. J. Azéma and M. Yor, Lecture Notes in Math., 1984, Springer. (§15)

[50] Abraham Pais, "Subtle is the Lord..." – The Science and the Life of Albert Einstein, 1982, Oxford University Press, New York. (§20)

[51] A.M. Perelomov, Generalized coherent states and some of their applications, Sov. Phys, Usp. *20,* 1977, 703-720. (§16)

[52] C. Piron, A unified concept of evolution in quantum mechanics, in Interpretations and Foundations of Quantum Theory, ed. H. Neumann, Grundlagen der exakten Naturwissenschaften *5,* 1979, 109-112. (§22)

[53] F. Rohrlich, Classical Charged Particles – Foundations of Their Theory, 1965, Addison-Wesley, Reading, Mass. (§22)

[54] P. Ruggiero and M. Zannetti, Critical phenomena at T = 0 and stochastic quantization, Phys. Rev. Letters *47,* 1981, 1231-1234.

[55] ————, Stochastic description of the quantum thermal mixture, Phys. Rev. Letters *48,* 1982, 963-966.

[56] ————, Quantum-classical crossover in critical dynamics, Phys. Rev. B *27,* 1983, 3001-3011.

[57] D.S. Shucker, Stochastic mechanics of systems with zero potential, J. of Functional Analysis *38,* 1980, 146-155. (§18)

[58] Lee Smolin, On the nature of quantum fluctuations and their relation to gravity and the principle of inertia, preprint, 1982, School of Natural Sciences, Institute for Advanced Study, Princeton, N. J. 08540, USA. (Current address: Dept. of Phys., Yale Univ., New Haven, Ct. 06520.) (§22)

[59] R.L. Stratonovich, Conditional Markov Processes and their Application to the Theory of Optimal Control, 1968, Amer. Elsevier, New York, (§8)

[60] Eugene P. Wigner, On hidden variables and quantum mechanical probabilities, in Quantum Mechanics, Determinism, Causality, and Particles, ed. M. Flato et al., 1976, 33-41, The Am. J. of Phys. (§23)

[61] D. Williams, Diffusions, Markov Processes, and Martingales, Vol. 1, 1979, Wiley, New York. (§4)

[62] Kunio Yasue, Stochastic calculus of variations, J. of Functional Analysis *41,* 1981, 327-340. (§14)

[63] ————, Quantization of dissipative dynamical systems, Phys. Letters, *64B,* 1976, 239-241.

[64] Kunio Yasue, Quantum mechanics of nonconservative systems, Annals of Phys. *114*, 1978, 479-496.

[65] Edward Nelson, Probability theory and Euclidean field theory, in Constructive Quantum Field Theory, ed. G. Velo and A. Wightman, Lecture Notes in Phys. *25*, 1973, Springer. (§23)

[66] D. Preiss and R. Kotecký, Markoff property of generalized random fields, Proceedings of the 7th Winter School of Abstract Analysis, Strážné, 1979. (§23)

[67] Eric Carlen, Conservative diffusions, Comm. in Math. Phys. *94*, 1984, 293-315. (§§11, 15)

[68] J.-C. Zambrini, Stochastic dynamics, Intern. J. of Theor. Phys. *24*, 1985. (§14)

[69] Lee Smolin, Derivation of quantum mechanics from a deterministic non-local hidden variable theory I. The two dimensional theory, preprint (see [58] for current address).

[70] P. Ruggiero and M. Zannetti, Microscopic derivation of the stochastic process for the quantum Brownian oscillator, Physical Review A, *28*, 1983, 987-993.

[71] W.A. Zheng and P.A. Meyer, Quelques resultats de "mecanique stochastique," to appear in Séminaire de Probabilités XVIII or XIX.

[72] Wolfgang Pauli, Pauli Lectures on Physics, Vol. 6, Selected Topics in Field Quantization, ed. C.P. Enz, MIT Press, Cambridge, Mass., 1973, Ch. 7. (§14)

[73] Bryce S. DeWitt, Dynamical theory in curved spaces I. A review of the classical and quantum action principles, Rev. Mod. Phys. *29*, 1957, 377-397. (§14)

[74] Francesco Guerra and Maria I. Loffredo, Thermal mixtures in stochastic mechanics, Lettere al Nuovo Cimento *30*, 1981, 81-87.

INDEX

LIST OF SYMBOLS

Library of Congress Cataloging in Publication Data

Nelson, Edward, 1932-
Quantum fluctuations.

Bibliography: p.
Includes index.
1. Quantum field theory. 2. Diffusion processes.
3. Stochastic processes. I. Title.
QC174.45.N45 1985 530.1'43 84-26449

ISBN 0-691-08378-9
ISBN 0-691-08379-7 (pbk.)

Milton Keynes UK
Ingram Content Group UK Ltd.
UKHW021914150524
442683UK00001B/72

9 780691 083797